Environmental Challenges and Solutions

Volume 4

Series editor
Robert J. Cabin, Brevard College, Brevard, NC, USA

Aims and Scope

The *Environmental Challenges and Solutions* series aims to improve our understanding of the Earth's most important environmental challenges, and how we might more effectively solve or at least mitigate these challenges. Books in this series focus on environmental challenges and solutions in particular geographic regions ranging from small to large spatial scales. These books provide multidisciplinary (technical, socioeconomic, political, etc.) analyses of their environmental challenges and the effectiveness of past and present efforts to address them. They conclude by offering holistic recommendations for more effectively solving these challenges now and into the future. All books are written in a concise and readable style, making them suitable for both specialists and non-specialists starting at first year graduate level.

Proposals for the book series can be sent to the Series Editor, Robert J. Cabin, at cabinrj@brevard.edu

More information about this series at http://www.springer.com/series/11763

John C. Pierce • Brent S. Steel

Prospects for Alternative Energy Development in the U.S. West

Tilting at Windmills?

Springer

John C. Pierce
School of Public Affairs and Administration
University of Kansas
Lawrence, KS, USA

Brent S. Steel
School of Public Policy
Oregon State University
Corvallis, OR, USA

ISSN 2214-2827 ISSN 2214-2835 (electronic)
Environmental Challenges and Solutions
ISBN 978-3-319-53413-8 ISBN 978-3-319-53414-5 (eBook)
DOI 10.1007/978-3-319-53414-5

Library of Congress Control Number: 2017931402

Printed on acid-free paper

This Springer imprint is published by Springer Nature
The registered company is Springer International Publishing AG
The registered company address is: Gewerbestrasse 11, 6330 Cham, Switzerland

Preface

The impact of humans on climate was first identified over 100 years ago by a Swedish scientist named Svante Arrhenius. Arrhenius' research was concerned with the adverse impact of burning coal on the atmosphere—more specifically the impact of CO2 emissions contributing to global warming. Increasing evidence that the use of fossil fuels was contributing to rising global temperatures led President Lyndon Johnson to issue a special message to Congress in 1965 that stated: "This generation has altered the composition of the atmosphere on a global scale through radioactive materials and a steady increase in carbon dioxide from the burning of fossil fuels." The mounting evidence of anthropogenic causes of climate change has led to over 97% of climate scientists concluding that humans are indeed negatively impacting global climate. However, at the time of publication of this book, there are approximately 182 climate deniers in Congress and the election of a new president who has consistently cast doubt on the science of climate change.

While the national government and many national political leaders have cast doubt on climate change and associated science, many state and local governments are taking the matter seriously. For example, the 2013 Pacific Coast Action Plan on Climate and Energy, signed by the Governors of California, Oregon, and Washington and the Premier of British Columbia, launched a broadly announced public commitment to reduce greenhouse gas emissions through multiple strategies. Those strategies included the development and increased use of renewable energy sources. The initiative recognized that citizens are both a central component in abating greenhouse gas emissions with regard to their energy use behaviors and are important participants in the public policymaking process at both state and local levels of government. However, with the promotion of alternative energy technologies also comes a variety of environmental tradeoffs—the subject of this book.

This work explores one of the striking anomalies in the development of renewable energy technologies. Those technologies have been developed largely in response to the twin challenges of finite energy resources and the climate change consequences of traditional carbon-based approaches to energy production. To be sure, there are a complex of factors leading to individuals' support for those alternative energy technologies (AETs), including such motives as economic incentives, depletion of natural resources, and concern for national security. At the same, though, by far the most powerful reason for individuals to support AETs seems to be their beliefs in negative environmental consequences of traditional energy technologies associated with carbon-based fuels, including global climate change.

The siting of AETs, such as wind, solar, geothermal, and wave energy technologies, often generates significant opposition from multiple sources. Ironically, as the case studies in the early chapters of this book demonstrate, a number of opponents to AET installations are environmentalists themselves. So, while our research establishes that in principle environmentalists are more likely than other citizens to support AET Installations, we are just as interested in what may distinguish pro-AET and anti-AET citizens. Relying on a series of five AET specific tradeoff questions designed specifically for this study, we are able to sort environmentalists into two groups labeled "globalists" and "localists." Those tradeoff questions were asked to nearly 3000 respondents in Washington, Oregon, California, and Idaho to indicate their level of support for an AET installation in the face of identified local negative environmental effects (e.g., installing a windmill farm when it disturbs rare tallgrass prairie or installing offshore wave energy turbines that may interfere with whale migration routes). The remaining chapters of the book focus on the relationship of the tradeoff beliefs to respondents' political values, political ideology, beliefs about energy scarcity, security and technology, knowledge levels, and information sources. The final chapter addresses the implications of the case studies and the survey-based chapters for participants in the AET installation processes.

Our hope is that this work contributes to the siting of AET facilities in an environmentally responsible manner that protects local environmental conditions and reduces global carbon emissions.

Acknowledgments

Brent Steel is grateful for the assistance and support of the following graduate students with the book: Mariana Amorim, Courtney Flathers, Andrew Spaeth, Lyndsay Trant, and Iaroslav Vugniavyi. He is espcially thankful for the assistance of Caryn Davis with editing and reorganization of the book.

John Pierce gratefully acknowledges the support, advice, and readiness to listen of the following faculty and graduate students: George Frederickson, John Nalbandian, Bonnie Johnson, Stacey White, Dorothy Daley, Rachel Krause, Holly Goerdel, Heather Getha-Taylor, Bill Budd, Jeannette Blackmar, Tony Reames, and Alisa Moldavanova.

The coauthors jointly wish to express their recognition of their decade-long valuable collaboration with Nicholas Lovrich, coauthor of many of our previous works and frequent participant and respondent in discussions about the direction of this work.

This research was generously supported by the School of Public Policy, Oregon State University, and by "Climate Change Adaptation, Sustainable Energy Development and Comparative Agricultural and Rural Policy," National Institute of Food and Agriculture (NIFA) *Higher Education Challenge (HEC) Grants Program.*

Contents

About the Authors

Brent S. Steel is professor of political science and director of the Oregon State University graduate program in public policy. He is the author, coauthor, or editor of numerous books and journal articles focusing on public values and beliefs in the area of environmental politics and policy, often in a cross-national context. Among his recent books are *Science in the Contemporary Policy Process: Responding to the Challenge of Complex Problems and Contested Policies* (Oregon State University Press, *forthcoming*), along with Edward Weber and Denise Lach; *Science and Politics: An A to Z Guide to Issues and Controversies* (Sage/*CQ* Press); *Acting as if Tomorrow Matters: Accelerating the Transition to Sustainability* (Environmental Law Institute, 2012), together with John Dernbach (lead author) et al.; and *State and Local Government: Sustainability in the 21st Century* (Oxford University Press, 2011), along with Christopher Simon and Nicholas P. Lovrich. Among Steel's recent scholarly articles are "An Assessment of the Impact that Auditory and Visual Experience with Wind Turbines Has on Support for Wind Production and Proximity-Based Opposition," *Society and Natural Resources* April (2016) DOI: 10.1080/08941920.2016.11719363, along with Rachel Krause and John C. Pierce; "Sustainable Futures: Healthy Lifestyles, Obesity and Access to Food in U.S. Counties 2012," *Agriculture and Agricultural Science Procedia* 8(2016):73–80, together with Erika Allen Wolters; "What Is the Best Available Science: A Comparison of Marine Scientists, Managers and Interest Groups," *Ocean and Coastal Management* 122(2016): 95–122, along with Erika Allen Wolters, Denise Lach, and Daniel Kloepfer; "Social Capital and the Centrality of Local Food Policy to Sustainability Initiatives in U.S. Counties," *Review of Studies of Sustainability* 3(2015): 45–62, along with Jeannette Blackmar and John C. Pierce; and "Environmental Value Considerations in Public Attitudes About Alternative Energy Development in Oregon and Washington," *Environmental Management* 55(2015): 634–645, together with John C. Pierce, Rebecca L. Warner, and Nicholas P. Lovrich.

John C. Pierce is lecturer and graduate faculty member in the School of Public Affairs and Administration at the University of Kansas and vice chancellor emeritus at the University of Colorado in Colorado Springs. He is the author, coauthor, or editor of numerous books and journal articles focusing on public values and beliefs in the area of environmental politics and policy, often in a cross-national context. Among his books are *Nice Trout* (Whitefish Press, Forthcoming); *Beginner's Creek*, collected personal essays (Cincinnati, OH: Whitefish Press, 2012); *River Earth: A Personal Map*, collected essays (Pullman, WA:WSU Press, 1999); *Political Culture and Public Policy in Canada and the United States: Only a Border Apart?* (Lewiston, NY: Mellen Press, 2000), along with Nicholas P. Lovrich, Jr., Brent S. Steel, Mary Ann E. Steger, and John Tennert; *Critical Masses: Citizen Responses to the Environmental Consequences of Nuclear Weapons Production in the United States and Russia* (Cambridge, Mass: MIT Press, 1999), together with Russell Dalton, Paula Garb, Nicholas P. Lovrich, Jr., and John Whitely; *Political Communication and Environmental Interest Groups: The United States and Canada* (Westport, CT: Praeger, 1992), along with Nicholas P. Lovrich, Jr., Mary Ann E. Steger, and Brent S. Steel; *Public Knowledge and Environmental Politics in Japan and the United States* (Boulder, CO: Westview, 1989), together with Nicholas P. Lovrich, Jr., Taketsugu Tsurutani, and Takematsu Abe; and *Water Politics, Democracy and the Technical Information Quandary* (Millwood, NY: Associated Faculty Press, 1986), along with Nicholas P. Lovrich, Jr. Pierce's recent scholarly articles have appeared in numerous journals, including *The Social Science Journal*, *Society and Natural Resources*, *Review of Studies of Sustainability*, *Environmental Management*, *Biomass and Bioenergy*, *Nonprofit Management and Leadership*, and *Sustainability*.

Part I
Introduction & Context

Chapter 1
Tradeoffs

Abstract This chapter provides an overview and introduction to the book, describing its purpose, and the "story" to be told in the subsequent chapters. The chapter begins with a brief discussion of climate change, followed by a short review of issues pertaining to the environmental hazards' consequences of current carbon-based technologies as drivers of alternative energy technology. The discussion then turns to the background to alternative energy development in the four case study states, which will include state energy profiles, renewable energy capacity, and emerging state renewable energy policies. The chapter will conclude with a brief summary of the alternative explanations for individuals' particular opinions about alternative energy technologies that will be investigated in the chapters that follow.

Keywords Alternative Energy Technology (AET) • State energy profiles • Renewable energy capacity • State energy policy

1.1 Introduction

> *President Barack Obama has championed the pollution-free energy, nearly doubling America's wind power in his first term as a way to tackle global warming. But all energy has costs, and the administration has been forced to accept the non-so-green sides of green energy as a means to an end....[This and other examples] highlight Obama's willingness to accept environmental* ***trade-offs****—pollution, loss of conservation land and the deaths of eagles—in hopes that green energy will help fight climate change"* (Cappiello 2013b, emphasis added).

> *For proponents of renewable energy, it seems like every time a technology is primed to save the world, it gets dogged by some ironic bit of environmental unpleasantness...The latest on this list of "dark green" technologies is thermal solar power, which is coming into vogue in the southern California desert...* (Lund 2014).

The growing world population and a rapidly increasing demand for energy are exerting enormous pressure on nature's limited resources. The satisfaction of this worldwide demand has relied heavily on fossil fuels—principally coal and oil—for liquid fuels and electricity production. The result has been that carbon dioxide and other greenhouse gas emissions from the combustion of fossil fuels have grown in volume at alarming rates. Likewise, other environmental problems also have been

J.C. Pierce, B.S. Steel, *Prospects for Alternative Energy Development in the U.S. West*, Environmental Challenges and Solutions 4,
DOI 10.1007/978-3-319-53414-5_1

linked directly to fossil fuel extraction and consumption, such as ash waste disposal, mining runoff, oil spills, air pollution, acid rain deposition, estuarial contamination and ocean pollution, and global climate changes.

Many international agreements and domestic policies have been developed to address these various problems. Moreover, countries around the world are intensifying the search for alternative fuels and renewable energies (e.g., wind, solar, and wave energy) that will be less damaging to the environment, less detrimental to human health, and accessible to local communities. Like some traditional sources of energy, renewable energy sources often can be tailored to the geographic characteristics of the area where they are produced. For example, regions with regular sun exposure could harness solar energy, windy areas could tap the power of the wind, coastal areas might obtain their electricity directly from the tides and waves of the ocean, and biofuels could be developed in renewable crop producing locations or in areas of forest biomass.

Although the siting and development of renewable, alternative energy technologies (AETs) can face significant obstacles in the form of start-up costs and technological capacity, support for AETs has occasionally emerged from unexpected sources, such as traditional, pro-development interests that have identified economic and employment gains to be made, especially in the context of government subsidies that might mitigate the cost of the AETS. At the same time, however, attempts to site AET projects have been met by intense opposition from a variety of sources, such as prevailing political and entrenched economic interests, as well as from many *environmentalists*—from whom one might expect support for non-carbon based renewable energy initiatives.

1.2 Environmental Tradeoffs

Conflicts between environmentalists and the developers of AETs have occurred in nearly every location where the environment provides opportunities for the creation of non-carbon based production of energy. Central to many such conflicts are what may be viewed as environmental “trade-offs” between the benefits of non-carbon based sources of energy and their perceived negative environmental consequences or potential harmful side effects. Examples of such tradeoffs include the effects of solar power on endangered turtles in the California desert or the impacts of geothermal energy production on subsurface configurations. Environmental advocates in Idaho and Wyoming may view wind farms as posing an excessive danger to birds or constituting aesthetic blots on the land. Likewise, hydropower may be seen as an ecosystem-altering cause of decline in Washington’s anadromous fisheries, Native American traditions and reservation viability, and wildlife habitat. Similarly, wave-energy farms may be opposed by environmentalists who characterize them as impediments to sea mammals and migrating fish, as well as disruptive of the ecology of the sea bottom off the Oregon Coast. Finally, some environmentalists may view biofuels generated from planted crops as inefficient, vis-à-vis energy inputs vs.

outputs; inequity-generating, with respect to increasing the cost of food, and destructive to grasslands.

Gaining a better understanding of such opposition to ongoing or proposed siting and development of AET projects is a central focus of this book. As noted, perhaps the most intriguing questions that have emerged through this study concern the increasing opposition to specific AET proposals from environmentalists, especially those with bio-centric (as opposed to anthropocentric) orientations to the environment. As Yonk et al. argue in the book *Green vs. Gree*n (2013, preface):

> The micro-goals of protecting individual areas, species, small-scale ecosystems, and other local environmental aims often limits ability to achieve macro-goals like preventing climate change of transitioning to large-scale green energy production.

But the questions we ask herein are more general than that, namely, do these anti-AET environmentalists differ from other environmentalists who support AETs, as well from non-environmentalists who support or oppose AETs? The book seeks to clarify whether identifiable individual-level attributes (e.g., values, interests, knowledge, demographic characteristics) lead to support for or opposition to the development and implementation of AETs. The book argues that there are indeed such discernible attributes, and, moreover, that the identification and exploration of those attributes are important for the development of support strategies for a well-informed and achievable set of policies regarding renewable sources of energy.

We explore these questions through a mixed-methods approach: the information and field data gathered for the book derive from a set of case studies focused on specific AET siting controversies in the western United States, stakeholder interviews, and opinion surveys of the general publics of California, Oregon, Washington, and Idaho. The case studies section begins with Chap. 2, which offers an overview of energy usage, renewable energy capacity, and renewable energy policy in California, Oregon, Washington, and Idaho. This chapter provides background for the case studies, which are presented in Chaps. 3, 4, 5, and 6. These case studies focus on specific controversies that illustrate the complexity surrounding the siting and development of AETs, including conflicts involving many different stakeholders. In the surveys section, consisting of Chaps. 7, 8, 9, 10, 11, and 12 we approach questions about citizens' beliefs and levels of knowledge about climate change, renewable energy policy, and AETs, as well as the information sources that inform their views. Finally, in Chap. 13 we review the lessons learned from this study and place them loosely within the context of the current U.S. political climate.

Thus, the book has two overall goals: The first is to identify empirically distinctive combinations of environmental and AET opinions, and then to describe how those types differ across such attributes as values, attitudes, demographic characteristics, scientific knowledge and geographic location. The second goal is to assess the implications of the study's findings for the development of appropriate strategies for the successful implementation of AETs in diverse socio-political settings—particularly in cases where local environmental tradeoffs may fuel controversy (see Box 1.1).

Box 1.1 Wind Farm Tradeoffs

Montana

"An operations center in California can shut down wind turbines 1200 miles away in Montana in fewer than 30 seconds when the flight patterns of golden eagles and other raptors indicate a potential collision in the making. The quick response is possible because of tracking radar, "detect and deter" cameras and human spotters called "bio-monitors" deployed at Rim Rock Wind Facility in Montana's Glacier and Toole counties. Greg Copeland of San Francisco-based NaturEner USA, the owner, said the combination of technology and trained avian biologists form a tiered alert system that's proving to be effective, so far, at preventing the turbine blades from striking birds"(Moscow-Pullman Daily News 2013).

Washington

"The government for the first time has enforced environmental laws protecting birds against wind energy facilities, winning a $1 million settlement Friday from a power company that pleaded guilty to killing 14 eagles and 149 other birds at two Wyoming wind farms. The Obama administration has championed pollution-free wind power and used the same law against oil companies and power companies for drowning and electrocuting birds. The case against the Duke Energy Corp. and its renewable energy arm was the first prosecuted under the Migratory Bird Treat Act against a wind energy company. A study in September by federal biologists found that wind turbines had killed at least 67 bald and golden eagles since 2008. Wyoming had the most eagle deaths. That did not include deaths at Altamont Pass, an area in Northern California where wind farms kill an estimated 60 eagles a year. Before this case, not a single wind energy company had been prosecuted for a death of an eagle or other protected bird, even though each death is a violation of federal law unless a company has a federal permit" (Cappiello 2013a).

Washington

"Under pressure from the wind-power industry, the Obama administration said Friday it will allow companies to kill or injure eagles without the fear of prosecution for up to three decades. The new rule is designed to address environmental consequences that stand in the way of the nation's wind energy rush: the dozens of bald and golden eagles being killed each year by the giant, spinning blades of wind turbines….President Barack Obama has championed the pollution-free energy, nearly doubling America's wind power in his first term as a way to tackle global warming. But all energy has costs, and the administration has been forced to accept the non-so-green sides of green energy as a means to an end….(This and other examples) highlight Obama's willingness to accept environmental trade-offs—pollution, loss of conservation land and the deaths of eagles—in hopes that green energy will help fight climate change" (Cappiello 2013b).

(continued)

Box 1.1 (continued)

California

For proponents of renewable energy, it seems like every time a technology is primed to save the world, it gets dogged by some ironic bit of environmental unpleasantness. Wind turbines can be dangerous to birds and bats. Biofuels require so much land, fertilizer, and water that they may use more energy to produce than they provide. Hydropower dams screw up a whole lot of river habitat and stop fish from spawning. The latest on this list of "dark green" technologies is thermal solar power, which is coming into vogue in the southern California desert... U.S. Fish and Wildlife employees and energy company staff found 233 birds of 71 different species at three California solar facilities—Ivanpah, Genesis, and Desert Sunlight—during random surveys over two years. That's not a huge number of birds (though the limited scope of the collections means it's just a fraction of the actual deaths), but what's shocking is the way some of these birds are dying: They are literally being burned alive, in midair (Lund 2014).

1.3 Environmental Values

Much of the book's foundation, based on survey research data, lies in a unique typology of citizens based in the combination of a measure of their core environmental values along with a composite measure of their responses to a series of trade-off questions that link the implementation of AETs with expected non-carbon related environmental or societal consequences. In particular, "globalists" (pro-environment and pro-AET) are compared with "localists" (pro-environment, anti-AET with negative environmental consequences). To explore different opinions regarding difficult environmental trade-offs involving renewable energy, the authors have developed an index based in individuals' responses to five questions such as the following: Would you rather "Develop high potential windmill energy 'farm' in a windy, rare 'tall grass prairie' area" — or would you rather "Preserve rare 'tall grass prairie" ecology even if this prevents windmill energy farm production."

In this way. The book examines the role of values in framing public orientations toward issues of renewable energy and the environment. In particular, Dunlap and Catton's development of the New Environmental Paradigm (now the "New Ecological Paradigm") is a way of representing value-based sets of orientations to the relationships between humans and the natural world (Dunlap and Catton 1979a, b; Dunlap et al. 2000). The increasing commitment to environmental values was broadly seen as growing out of the postindustrial society of the 1960s and beyond, a period of cultural change produced by Western societies enjoying general peace and a prolonged period of prosperity (Inglehart 1991, 1997). The impact of New Environmental Paradigm values and of what Inglehart called "Post-materialist" values (as opposed to "materialist" values) has been widely examined by numerous scholars in several countries featuring distinctive political and social histories and cultures such as the United States, Japan, Russia, and Canada (Dalton et al. 1999; Pierce et al. 1987, 1992, 2000).

Viewed from one perspective, AETs can be seen as a product of the growth of this environmental consciousness, including concerns about global warming and regional climate change. These and public health-related environmental concerns have been widely identified as central to cultural change in the 1960s and 1970s (e.g., Inglehart 1991, 1997). The goal of achieving global sustainability implies a capacity to substitute renewable energy sources for nonrenewable, finite sources of energy such as coal and oil, the by-products of which are potentially hazardous to humans and the environment. Thus, one would expect support for these alternative energy technologies clearly to be found among individuals professing environmental values, and among those persons holding general political values associated with environmentalism.

Viewed from the perspective of prevailing interests and established patterns of economic life, many interested parties may see no need for the development of these renewable resources with new, unfamiliar forms of technology and the new types of government investment incentives associated with them (Steel et al. 2015). Furthermore, oppositional forces may raise doubts as to the validity of claims of global climate change science, similarly cast a cloud of doubt on the assumption of a rapidly diminishing carbon-based fuel supply in light of the great abundance of "clean coal" in North America, and tend to dispute the severity of the environmental consequences of contemporary means of energy production oriented toward fossil fuels. These views are disseminated through a broad array of information and sometimes disinformation sources. Thus, a better understanding of not only what citizens may believe about climate change and renewable energy technologies, but where they turn for information is vitally important in shaping future efforts to successfully site and develop AETs. Finally, understanding the dynamics of the sometimes seemingly improbable marriage of values and policy will also be crucial to identifying those strategies most likely to be successful in implementing alternative energy technologies.

References

Cappiello, D. (2013a, November 22). In a first, wind utility hit for killing eagles. *Los Angeles Daily News*. http://www.dailynews.com/environment-and-nature/20131122/wind-farm-gets-fined-for-killing-eagles. Accessed 15 Aug 2016.

Cappiello, D. (2013b, December 6). U.S. to allow eagle deaths to aid wind power. *Daily News*. http://www.nydailynews.com/news/national/u-s-killing-eagles-aid-wind-power-article-1.1540116. Accessed 15 Aug 2016.

Dalton, R. J., Garb, P., Lovrich, N. P., Pierce, J. C., & Whiteley, J. M. (1999). *Critical masses: Citizens, nuclear weapons production, and environmental destruction in the United States and Russia*. Cambridge, MA: MIT Press.

Dunlap, R. E., & Catton Jr., W. R. (1979a). Environmental sociology. *Annual Review of Sociology, 5*, 243–273.

Dunlap, R. E., & Catton Jr., W. R. (1979b). Environmental sociology: A framework for analysis. In T. O'Riordan, R. Kerry Turner, & R. C. d'Arge (Eds.), *Progress in resource management and planning* (pp. 57–85). Chichester: Wiley.

Dunlap, R. E., Van Liere, K., Mertig, A., & Jones, R. (2000). Measuring endorsement of the new environmental paradigm: A revised NEP scale. *Journal of Social Issues, 56*, 425–442.

Inglehart, R. (1991). *Culture shift in postindustrial society*. Princeton: Princeton University Press.

Inglehart, R. (1997). *Modernization and postmodernization: Cultural, economic, and political change in 43 societies*. Princeton: Princeton University Press.

Lund, N. (2014, July 8). Solar power plants are literally roasting birds in-flight. *Slate*. http://www.slate.com/blogs/wild_things/2014/07/08/solar_power_plants_burning_birds_usfws_report_on_california_facilities_and.html. Accessed 15 Aug 2016.

Moscow-Pullman Daily News. (2013, November 9). Wind farm tries to minimize bird deaths. *Moscow-Pullman Daily News*. http://dnews.com/news_ap/article_42f06690-2255-548f-a4ec-625c2099b13e.html. Accessed 15 Aug 2016.

Pierce, J. C., Lovrich, N. P., Tsurutani, T., & Abe, T. (1987). Environmental belief systems among Japanese and American elites and publics. *Political Behavior, 9*, 139–159.

Pierce, J. C., Steger, M. A., Steel, B. S., & Lovrich, N. P. (1992). *Citizens, political communication, and interest groups: Environmental organizations in Canada and the United States*. New York: Praeger Publishers.

Pierce, J. C., Lovrich, N. P., Steel, B. S., & Steger, M. A. (2000). *Political culture and public policy in Canada and the United States: Only a border apart?* New York: Edwin Mellen Publishers.

Steel, B. S., Pierce, J. C., Warner, R. L., & Lovrich, N. P. (2015). Environmental value considerations in public attitudes about alternative energy development in Oregon and Washington. *Environmental Management, 55*, 634–645.

Yonk, R., Simmons, R. T., & Steed, B. C. (2013). *Green vs. Green: The political, legal and administrative pitfalls facing green energy production*. New York: Routledge.

Part II
Case Studies

Chapter 2
Background: Energy Use, Capacity, and Policies

Abstract This chapter provides background and context for the book, including the four alternative energy case studies provided in Part II and the analysis of public opinion data in Part III. The development of alternative energy technologies is a primary focus of energy policy in California, Oregon, and Washington, with each of the states engaged in regional efforts with British Columbia to promote the Pacific Coast Action Plan on Climate and Energy of 2013. Although more conservative Idaho is not a signatory to the action plan, it does have enormous renewable energy capacity and does offer some renewable energy incentive programs in conjunction with utilities. This chapter provides each state's energy profile in terms of overall consumption and the major sources of energy currently in use, followed by an overview of state renewable energy capacity, as well as current regional and state energy policies.

Keywords Pacific Coast Action Plan on Climate and Energy • State energy use • State energy policy • State energy capacity

2.1 Introduction

> *With responsible energy development, resource exploration in the Arctic will provide more reliable and affordable energy and an opportunity for the United States to grow jobs and enhance its national security. Energy development is crucial for the future of the West Coast and the pipeline. The U.S. should be a leader, not a follower, in establishing sustainable development of Arctic [petroleum] energy – for its own energy security and national security* ("Why Oregonians should support Arctic oil extraction" David Holt, *The Oregonian*, April 13, 2014).

> *California, Oregon and Washington could cut their petroleum use by half in the next 15 years through policies that encourage greater transportation options and the more robust use of existing and emerging low-carbon technologies… [and these states] could serve as powerful models for reducing petroleum consumption as the world confronts the reality that fossil fuel consumption is leading to dangerous climate change* ("West Coast States Could Cut Petroleum Consumption in Half by 2030" Union of Concerned Scientists, January 28, 2016).

J.C. Pierce, B.S. Steel, *Prospects for Alternative Energy Development in the U.S. West*, Environmental Challenges and Solutions 4,
DOI 10.1007/978-3-319-53414-5_2

The four states of California, Oregon, Washington, and Idaho create a strikingly important context in which to focus this study. The September 2003 West Coast Governors' Global Warming Initiative and the subsequent 2013 Pacific Coast Action Plan on Climate and Energy signed by the Governors of California, Oregon and Washington and the Premier of British Columbia, launched a broadly announced public commitment to reduce greenhouse gas emissions through multiple strategies. Those strategies include the development and increased use of renewable energy sources. The initiative recognized that citizens are both a central component to abating greenhouse gas emissions with regard to their energy use behaviors, and are important participants in the public policymaking process at both state and local levels of government. These signatories also are states whose respective governors have been active participants in the Governors' Global Climate Summits co-hosted by the United Nations Environment Programme (UNEP).

Moreover, the three states are also host to numerous alternative energy technology (AET) installation controversies. Likewise, each state has a long history of environmental concerns, and each has spawned major home-grown environmental movements; in addition, each state has extensive experience with both carbon-based and non-carbon-based (such as hydroelectric) energy production for the power grid. Idaho also is included as a comparison due to its western geographical location bordering Washington and Oregon, but also because of its much less aggressive policy approach in adopting and promoting renewable energy technologies, its historically more conservative politics and culture, and its non-signatory participation in the coastal states' action plan.

In order to provide some context for the case studies and the surveys, this chapter provides each state's energy profile in terms of overall consumption and the major sources of energy currently in use. This background information is followed by a brief review of regional and state renewable energy capacity, as well as current regional and state energy policies. The energy policy environment is dynamic, with states making frequent changes and adjustments to existing policies. Thus, the review of policies must be general in order to illustrate state commitment to renewable energy development (see Appendix A).

2.2 California Energy Usage

With a population of 38.3 million and the largest state economy in the country, California uses a significant amount of energy (U.S. Department of Energy 2012a). In 2012, California's total energy consumption was 7641 trillion British thermal units (Btu), which was second only to Texas. Despite being #2 in *per capita* total energy consumption at 209 Btu, California had one of the lowest *per capita* energy consumption levels in the U.S., ranking 47th nationally, in 2011 (U.S. Department of Energy 2012a). California's total energy expenditures were second in the country at $136,364 million, while expenditures per capita (2011) were ranked 47th at $3612 (U.S. Department of Energy 2014). In 2014, California's labor force was estimated at 18.6 million, with a per capita income of $44,980. Major contributors to California's low per capita energy use are the state's mild/dry climate and successful energy efficiency programs (U.S. Department of Energy 2014). Moreover, abundant energy resources are available to the state.

The United States Department of Energy monitors four sectors of energy consumption: residential, commercial, industrial, and transportation. The California residential sector had a 2012 end-use consumption total of 1472 trillion Btu's, ranking second in the nation with a 7.4% share of the national total. Residential end-use expenditures were $19,212 million in 2012 for an 8.2% share of the U.S. total (U.S. Department of Energy 2012a). Despite having the largest population, California residential energy use *per capita* was 5th lowest in the nation, at 209 Btu (U.S. Department of Energy 2014). California homes were also among the lowest in the nation in average site electricity consumption, at 6.9 megawatt hours per year (U.S. Department of Energy 2012b). Almost two-fifths of Californians reported having no air conditioning system or not using their air conditioning regularly. This relatively moderate air conditioning usage pattern is due in large part to the mild, dry climate of California's more populated urban centers.

Renewable resources (excluding hydroelectric power) accounted for 24.7% of net electricity generation in California, three times the national average of 7.8%.

California's commercial sector had an end-use energy consumption total of 1481 trillion Btu in 2012, accounting for 19.4% of California's energy consumption and 8.5% of the U.S. total for that category (U.S. Department of Energy 2012a). Expenditures in this sector were $18,967 million, for a national share of 11%. Commercial energy prices in 2012 were $26.33 per million Btu (U.S. Department of Energy 2012c).

The industrial sector in California had an end-use consumption total of 1744 trillion Btu in 2012, accounting for 5.6% of the national total. Even though California's industrial end-use consumption ranked second nationally, the industrial sector had the lowest total expenditures, $13,029 million or $12.68 per million Btu; this accounted for 22.8% of California's total energy consumption and 5.8% of the U.S. total (U.S. Department of Energy 2012a).

In 2012, end-use energy consumption in California was dominated by the transportation sector, which used 2943 trillion Btu and accounted for 11.0% of U.S. consumption in this sector (U.S. Department of Energy 2012a). Expenditures in this sector were $85,157 million or 11.8% of the U.S. share, which was the largest amount among all states. The California transportation sector is dominated by a number of major airports, military bases, the largest number of registered vehicles of any state, and some of the longest commute times in the U.S (U.S. Department of Energy 2014). In 2012, California drivers traveled an estimated 326,272 million miles. Energy demand in this sector led California to adopt several regulations aimed at minimizing air pollution and environmental damage. As of 2012, California had the largest number of registered alternative fuel vehicles in the nation. Furthermore, the state requires all motorists to use either California Reformulated Gasoline (CaRFG) or California Oxygenated Reformulated Gasoline, special blends designed to reduce smog-forming emissions from motor vehicles (California EPA 2014).

In 2012, California end-use consumption totaled 620.2 million barrels of oil (9.2% of the U.S. total) with expenditures of $91,223 million (10.3% of U.S. total); 2403.4 billion cubic feet of natural gas (9.4% of the U.S. total) with expenditures of $13,009 million (9.8% of U.S. total); and 1.9 million short tons of coal (0.2% of the U.S. total) with expenditures of $134 million (0.3% of U.S. total) (U.S. Department of Energy 2012b). Sources consumed for electricity generation included petroleum, natural gas, and coal. California consumption of natural gas for electricity generation totaled 60,310 million cubic feet, for a U.S. share of 10.3%. The amount of coal used in that state was 35 thousand short tons, which was an insignificant amount of the national total and not reported (U.S. Department of Energy 2014).

California is very energy rich. In 2012, California produced 2335 trillion Btu, tenth nationally (U.S. Department of Energy 2012d). Despite a decline in oil production over the last two decades, California is still a leader in crude oil production. In 2013, California ranked third nationally in crude oil production and third in refining capacity. There were 18 refineries operating within the state capable of producing 2 million barrels of oil each day, or 7% of the U.S. total (U.S. Department of Energy 2012a). As of March 2014, crude oil production was 17,368 thousand barrels, which was 6.8% of total U.S. production. At the end of 2012, California had 2974 million barrels of crude oil in reserve, accounting for 9.7% of U.S. crude oil reserves (U.S. Department of Energy 2014).

Renewable energy sources were also significant sources of total energy production. In 2013, California ranked fourth in production of hydroelectric power, first in production of geothermal power, and second in production from other sources of renewable energy. Despite the permanent shutdown of the San Onofre Nuclear Generating Station, California ranked fifteenth nationally in nuclear power production, generating 1626 thousand Megawatt hours in 2013. Nuclear power generation totaled 6.8% of total state energy production in 2013, which was well below the national average of 18.8% for nuclear energy (U.S. Department of Energy 2014).

In 2012, 60% of California's net electricity generation came from natural gas-fired power stations, well above the U.S. average of 23.4%. Renewable resources (excluding hydroelectric power) accounted for 24.7% of net electricity generation in California, three times the national average of 7.8%. Hydroelectric power, at 6.9% of the state total, and nuclear power, at 6.8%, were other sources of non-carbon based power generation (U.S. Department of Energy 2014).

2.3 Oregon Energy Usage

As of 2013, Oregon had a population of approximately 3.9 million people, with a total energy consumption of 986 trillion Btu, which ranked 32nd in the nation. The state civilian workforce was 1.9 million (May 2014), with a per capita income of $40,223 (U.S. Department of Energy 2014). Total energy consumption per capita was ranked 39th in the nation at 253 million Btu. Total energy expenditures were $14,918 million in 2012, ranking 31st nationally (U.S. Department of Energy 2012a). A majority of Oregonians lives within the mild weather regions of the Pacific Coast or the temperate Willamette Valley, which is the primary reason for Oregon's low energy consumption totals. The state has an abundance of renewable energy resources, and in recent years, there has been an increased focus on the development of geothermal and wave energy technologies.

Oregon's residential sector had a 2012 end-use consumption total of 246 trillion Btu, ranking 27th in the U.S., with 1.2% of the total. Residential end-use expenditures were $2548 million in 2012 for a 1.1% share of the national total (U.S. Department of Energy 2012a). Energy use per capita was one of the lowest in the nation at 253 Btu, ranking 39th (U.S. Department of Energy, 2012a). Residential electricity prices were $21.52 per million Btu, 15.76% below the national average in April 2014 (U.S. Department of Energy 2014). Heating was the primary use of electricity in the residential sector, with close to 50% of Oregon homes reporting the use of electric heaters. Another 33% used natural gas, while the remaining homes used heating oil and wood (U.S. Department of Energy 2012f).

In 2012, Oregon's commercial sector had an end-use total of 188 trillion Btu, ranking 29th in the U.S., while accounting for 1.1% of the U.S. total. Commercial end-use expenditures were $1662 million, for a national share of 1% (U.S. Department of Energy 2012a). Commercial energy prices in 2012 were $18.92 per million Btu, 28th nationally. In April 2014, energy prices in the commercial sector were 15.29% below the national average (U.S. Department of Energy 2012f).

The industrial sector had an end-use consumption total of 245 trillion Btu in 2012, ranking 34th nationally with a share of 0.8% (U.S. Department of Energy 2012a). The industrial sector had total expenditures of $1794 million, accounting for 0.8% of the national total. Prices for this sector were $11.12 per million Btu, ranking 38th in 2012. In April 2014, energy prices for the industrial sector were 13.94% below the national average (U.S. Department of Energy 2014).

In 2012, the transportation sector had the highest energy consumption in Oregon, with an end-use total of 307 trillion Btu, 1.1% of the national total. Expenditures for this sector were $8913 million or 1.2% of the national total (U.S. Department of Energy 2012a). Prices for the transportation sector were the 5th highest in the U.S. at $29.53, which is 5.9% above the national average (U.S. Department of Energy 2012c). Vehicles traveled 33,173 million miles in 2012, which ranked 30th nationally (U.S. Department of Energy 2012a).

Sources consumed for electricity generation in Oregon included petroleum, natural gas, and coal. In 2012, end-use consumption totaled 63.3 million barrels of oil,

0.9% of the U.S. total, with expenditures of $9831 million, 1.1% of U.S. total; 215.8 billion cubic feet of natural gas, 0.8% of U.S. total, with expenditures of $1338 million, 1.0% of U.S. total; and 1.7 million short tons (2000 pounds each) of coal, 0.1% of U.S. total with expenditures of $55 million, 0.1% of U.S. total (U.S. Department of Energy 2012a). Consumption of natural gas for electricity generation totaled 2556 million cubic feet, for a U.S. share of 0.4%. Coal was also used in electricity generation, with 10,000 short tons consumed (U.S. Department of Energy 2012e).

In 2012, Oregon had total energy production of 497 trillion Btu, 30th nationally. Oregon has no oil reserves, no refining capacity, and no oil production. More than 90% of refined petroleum products in Oregon come from Washington refineries (U.S. Department of Energy 2012a). Natural gas production in 2012 was 770 million cubic feet and came from The Mist gas field, the only natural gas field used for production in the Pacific Northwest. Natural gas stores are held in seven underground storage tanks that have a total capacity of 29 billion cubic feet (U.S. Department of Energy 2012a).

Total net electricity generation was 4650 thousand MWh for the period ending April 2014. A significant portion of this, approximately 75%, came from hydroelectric dams. In April 2013, Oregon generated 3702 thousand MWh of hydroelectric power, which was 8.4% of the national total (U.S. Department of Energy 2012a). In years with an average rainfall, approximately half of Oregon's total power generation is in the form of hydroelectric power. In wetter years, this total can jump to over 70% (U.S. Department of Energy 2012a). In 2013, Oregon was second only to Washington in total hydroelectric power generation (U.S. Department of Energy 2014).

Other renewable resources are also used in energy production in Oregon. Wind energy facilities are found in the hills of eastern Oregon and the Columbia Gorge and the state has over 3000 megawatts of active wind farms. In geothermal energy production, Oregon ranked 3rd in the U.S. in 2013. Several geothermal facilities were operational in 2013 with a combined generating capacity of 2200 megawatts of electricity (U.S. Department of Energy 2014).

Other sources of energy in Oregon include wood/wood waste, which accounted for 1.1% of Oregon's total energy production; coal-fired power, 0.3% of the state total; biofuels, with 5.5 trillion Btu of energy production; and petroleum-fired which was an insignificant total (U.S. Department of Energy 2012a & 2012e). Oregon is also in the process of developing wave energy technologies with several projects taking place off the Oregon coast. The state has no nuclear generating capacity and little solar generating capacity at this time (U.S. Department of Energy 2012a & 2012c).

2.4 Washington Energy Usage

As of 2013, Washington was the 13th most populated state, with approximately 6.6 million people, a civilian labor force of approximately 3.5 million, and a per-capita income of $47,031 (U.S. Department of Energy 2014). The gross domestic product of the state was $375.7 billion, 12th nationally, and was led by the manufacturing sector. Total energy consumption in 2012 was 2057 Btu, ranked 16th in the nation, yet per capita consumption was ranked 27th in the nation at 298 million Btu (U.S. Department of Energy 2014). In 2014, Washington had total energy expenditures of $27,570 million, 16th in the nation, while per-capita expenditures were among the bottom third in the nation, 36th, at $3998, due in large part to mild weather in the more densely populated areas of the state. Residential energy consumption in Washington accounts for approximately 25% of total consumption, which helps keep per-capita costs low. Washington also has an abundance of renewable energy resources, particularly hydroelectric power, and leads the nation in electricity generated from such sources (U.S. Department of Energy 2014).

In 2012, Washington's residential sector had an end-use consumption total of 480 trillion Btu, which accounted for 2.4% of the U.S. total. End-use expenditures for the residential sector were $4396 million in 2012 and accounted for 1.9% of the national total (U.S. Department of Energy 2014). The residential sector used the second lowest amount of energy among all sectors in Washington in 2012, primarily due to a mild climate marked by cool summers and mild winters (U.S. Department of Energy 2012g). This allowed the state in May 2014 to boast the lowest average retail price of electricity to ultimate consumers, 8.93 cents per kilowatt-hour. For 2013, Washington had the lowest energy prices for residential consumers in the U.S. (U.S. Department of Energy 2014).

In 2012, Washington's commercial sector had an end-use total of 375 trillion Btu, the lowest of all sectors, and accounted for 2.2% of the U.S. total (U.S. Department of Energy 2014). Total expenditures for this sector were $3068 million, for a national share of 1.8%. Commercial energy prices in 2012 were $18.33 per million Btu (U.S. Department of Energy 2012c). In April 2014, commercial electricity prices were 22.31% lower than the national average (U.S. Department of Energy 2014).

The industrial sector in Washington had an end-use consumption total of 582 trillion Btu in 2012, which was 1.9% of the national total (U.S. Department of Energy 2014). Industrial end-use consumption ranked second in Washington, but had the lowest total expenditures, $2790 million or $9.72 per million Btu (U.S. Department of Energy 2012c). Electricity for the industrial sector was among the lowest in the nation in April 2014, with prices 37.63% lower than the national average (U.S. Department of Energy 2014).

In 2012, the transportation sector had the highest end-use consumption in Washington, using 620 trillion Btu, or 2.3% of the national total. Expenditures for this sector were over four times greater than the next highest sector, coming in at $17,316 million or 2.4% of the U.S. total. Washington drivers traveled 56,762 million miles in 2012, which was 20th in the nation (U.S. Department of Energy 2014). The price of energy in this sector was impacted by the mandatory use of oxygenated gasoline in motor vehicles, which required special refineries. Energy prices in this sector were $28.36 per million Btu, which was 1.6% higher than the national average in 2012 (U.S. Department of Energy 2012c).

In 2012, Washington consumed 139.2 million barrels of oil, or 2.1% share of U.S. total; 264.3 billion cubic feet of natural gas, or 1.0% of U.S. total; and 2.6 million short tons of coal, or 0.3% of U.S. total (U.S. Department of Energy 2014). Consumption for electricity generation totaled 2000 barrels of petroleum, 2200 million cubic feet of natural gas, and 129 thousand short tons of coal. For home heating, 35.5% of consumption was from natural gas, 2.9% from fuel oil, 53.0% from electricity, 3.2% from liquefied petroleum products, and 5.4% from other sources. In overall consumption, Washington consumed 42.8 trillion Btu of coal; 271.7 trillion Btu of natural gas, 301.3 trillion Btu of gasoline (excluding ethanol); 137.7 trillion Btu of distillate fuel; 63.3 trillion Btu of residential oil; 115.1 trillion Btu of other petroleum products; 97.8 Btu from nuclear power; 851.3 trillion Btu from hydroelectric power; 123.2 trillion Btu from biomass sources; 64.5 trillion Btu from other renewable sources; and 16.2 trillion Btu from LPG. Expenditures totaled $18,858 million for petroleum products; $2261 million for natural gas; and $90 million for coal (U.S. Department of Energy 2014).

In 2012, Washington produced 1110 trillion Btu, which was ranked 17th nationally (U.S. Department of Energy 2012d). A majority of Washington's energy production came from hydroelectric power. Washington is the largest producer of hydroelectric power in the United States, producing 29% of the nation's total (U.S. Department of Energy 2012d). Ten of the nation's 12 largest hydroelectric dams can be found in Washington, including the largest, Grand Coulee Dam, which is capable of producing 6809 megawatts of electricity. Over 77% of Washington's electricity generation came from hydroelectric power in 2013, which was more than nine times greater than the national average of 8.4% (U.S. Department of Energy 2014).

Despite having no crude oil production, Washington had the 5th highest crude oil refining capacity. In 2013, Washington had five fully operational refineries with a crude refining capacity of 631,700 barrels per calendar day, approximately 1.4% of the U.S. total. As of April 2014, Washington had stocks of 1303 thousand barrels of motor gasoline; 1773 thousand barrels of distillate fuel oil; and 30,194 million cubic feet of natural gas (underground storage) (U.S. Department of Energy 2014).

Beyond hydroelectric power, Washington also depends on other renewable resources for energy production. Washington was 10th nationally in wind power generation in 2013 (U.S. Department of Energy 2014); other renewable energy sources accounted for 9.2% of energy production for 2013. Natural gas, 2.9%, coal-fired, 1.9%, and nuclear, 8.2%, were other sources of power generation (U.S. Department of Energy 2014).

2.5 Idaho Energy Usage

In 2012, Idaho had a population of approximately 1.6 million and the lowest average per capita income in the United States at $35,382 (U.S. Census Bureau). Idaho's total energy production was 42nd in the nation in 2012, with 158 trillion Btu (U.S. Department of Energy 2014). Idaho also was 42nd in total consumption at 519 trillion Btu, with a total consumption per capita that was 21nd in the nation, at 325 million Btu (U.S. Department of Energy 2014). This consumption level is due in large part to the wide range in temperatures across the state. Harsh winters and hot summers increase demand for energy, as evident by the gap between production (158 trillion Btu) and consumption (519 trillion Btu) (U.S. Department of Energy 2012e). Despite this fact, Idaho had the 4th lowest average electricity price in the nation (U.S. Department of Energy 2014). Idaho's total energy expenditures were $6725 million, good for 43rd nationally, while total expenditures per capita were $4215, placing it 32nd in the nation (U.S. Department of Energy 2014). It is important to note that approximately 75% of Idaho's energy comes from out of state, but 78% of the energy production within the state comes from renewable sources (primarily hydroelectric), which is the highest in the nation (U.S. Department of Energy 2012e).

Idaho's residential sector had a 2012 end-use consumption total of 116 trillion Btu, which was 36th in the nation at 0.5% of the national total. Residential end-use expenditures were $1023 million in 2012, for a 0.4% share of the U.S. total (U.S. Department of Energy 2014).

The commercial sector of Idaho had an end-use total of 83 trillion Btu in 2012, accounting for 0.5% of the U.S. total (U.S. Department of Energy 2014). Expenditures in this sector were $625 million, for a national share of 0.4%. Commercial energy prices in 2012 were $15.37 per million Btu and were 2nd lowest in the nation, behind Illinois (U.S. Department of Energy 2012c).

In 2012, end-use energy consumption was led by the industrial sector, which consumed 36% of Idaho's energy. Idaho's industrial sector had an end-use consumption total of 187 trillion Btu in 2012, accounting for 0.6% of the U.S. total. Expenditures for this sector totaled $1311 million or $11.59 per million Btu, accounting for 0.6% of the U.S. total (U.S. Department of Energy 2014). The industrial sector is dominated by agriculture, forest products, and mining, as well as by the emerging industries of electronics manufacturing and tourism (U.S. Department of Energy 2012e).

The Transportation sector consumed 133 trillion Btu, 0.5% of the national total (U.S. Department of Energy 2014). Expenditures in this sector totaled $3766 million, 0.5% of the national total, which was among the smallest amounts in the country. The price per Btu for the transportation sector was 3rd highest in the nation, at $29.64 (U.S. Department of Energy 2012a).

In 2012, petroleum end-use consumption totaled 29.8 million barrels, 0.4% of the national total, with expenditures of $4506 million, 0.5% of the national total; 89.0 billion cubic feet of natural gas, 0.3% of the U.S. total with expenditures of

$544 million, 0.4% of U.S. total; and 0.3 million short tons of coal with expenditures of $16 million. Sources consumed for electricity generation included petroleum, natural gas, coal, and a variety of renewable energy sources. Idahoans consumed 104.1 trillion Btu of hydroelectric energy, 34.4 trillion Btu of biomass-generated energy, and 20.3 trillion Btu of power from other renewable energy sources, including wind and geothermal.

Idaho is rich in renewable energy resources, which include hydroelectric, wind, solar, and geothermal power. In 2012, Idaho produced 158 trillion Btu, ranked 47th nationally (U.S. Department of Energy 2014). In 2013, approximately 78% of Idaho's energy production came from hydroelectric and other renewable sources, with 16% alone coming from wind generation (an increase of over 35% over 2012). The largest source of energy was hydroelectric generation, which accounted for 58% of the net total of energy produced in Idaho in 2013. Energy generation from biofuels totaled 6.8 trillion Btu in 2013 (U.S. Department of Energy 2014).

Idaho does not produce coal, natural gas, or petroleum. Exploration for oil reserves in Idaho has taken place for over 100 years, yet no reserves have been discovered. Petroleum products are delivered to Idaho via two pipelines, one from Utah and one from Montana (U.S. Department of Energy 2012e). Natural gas fields have been discovered in Idaho and are currently in development. Natural gas supplies are presently transported by pipeline from Canada and other western states. Coal reserves are minimal and are generally considered non-recoverable. The state has no power plants that operate solely on coal. There are two cogeneration plants (the simultaneous production of heat and energy) that produce coal-fired energy. Coal for these plants is mined in Utah and Wyoming and transported to Idaho by train (U.S. Department of Energy 2012e).

2.6 Regional Renewable Energy Capacity

In terms of overall renewable energy *capacity*, California, Oregon and Washington are leading producers in the western U.S. The data displayed in Table 2.1 show that the three states were ranked highest out of all western states in installed renewable energy capacity in 2013, with California capable of producing 15,534 MW, Oregon 3652 MG, and Washington 3239 MW. Similarly, the three states were ranked in the top four of western states for actual renewable fuels production, with California first at 292 mGy, Oregon second at 167 mGy, and Washington fourth at 120 mGy.

When examining specific sources of installed renewable energy, California, Oregon, and Washington have substantial wind power and biomass capacity (see Table 2.2). In 2013 California had 5829 MW wind power capacity, Oregon had 3153 MW, and Washington had 2808 MW. Idaho was at a much lower level of 973 MW capacity. When examining other sources of renewable energy capacity, it is apparent that California has a very robust capacity in solar photovoltaic, solar thermal electric, geothermal power, biomass power, ethanol, and biodiesel. This is partly because of available resources and the size of the state in terms of population and geography, but also because of the state's very aggressive renewable energy

Table 2.1 Installed renewable capacity of western states—2013

Rank	Renewable power (not including hydro)[a] (MW)	Renewable fuels (mGy)
1	California: 15,534	California: 292
2	Oregon: 3652	Oregon: 167
3	Washington: 3239	Colorado: 125
4	Colorado: 2715	Washington: 120
5	Arizona: 2126	Arizona: 83
6	Wyoming: 1411	Idaho: 55
7	Nevada: 1209	New Mexico: 32
8	Idaho: 1139	Utah: 22
9	New Mexico: 1045	Nevada: 16
10	Hawaii: 831	Wyoming: 10
11	Montana: 648	Hawaii: 6
12	Utah: 428	Alaska: 0.3
13	Alaska: 216	Montana: 0.3

Source: American Council on Renewable Energy 2014
[a]MW = megawatt; mGy = million gallons per year

Table 2.2 Installed renewable energy—2013

Source:	California	Idaho	Oregon	Washington
Wind Power	5829 MW[a]	973 MW	3153 MW	2808 MW
Solar Photovoltaic	5183 MW	2 MW	63 MW	27 MW
Solar Thermal Electric	393 MW	0 MW	0 MW	0 MW
Geothermal Power	2711 MW	16 MW	33 MW	0 MW
Biomass Power	1417 MW	148 MW	403 MW	403 MW
Ethanol	223 mGy[a]	50 mGy	149 mGy	0 mGy
Biodiesel	70 mGy	5 mGy	18 mGy	120 mGy

Source: American Council on Renewable Energy (2014)
[a]*MW* megawatt, *mGy* million gallons per year

policies, which will be discussed below. Oregon also has some capacity in solar photovoltaic, geothermal power, and ethanol. Washington also has some solar photovoltaic capacity, as well as enormous biodiesel capacity, compared with the other three states.

2.7 Regional Renewable Energy Policies

Before providing brief overviews of each state's renewable energy policies, it is important to identify the regional commitment to renewable energy policy and climate change mitigation. In October, 2013 the governors of California, Oregon, and Washington, along with the Premier of British Columbia, signed the Pacific Coast Action Plan on Climate and Energy. As the preamble of the agreement states, the nonbinding plan commits each political jurisdiction to reduce greenhouse gases, support and encourage renewable energy sources, provide clean transportation, develop energy efficiencies, and cooperate at the local, regional and international levels on climate and energy policies (Box 2.1).

In a joint statement concerning the plan, the three governors and the premier announced (Brown et al. 2014):

Box 2.1 Pacific Coast Action Plan on Climate and Energy

Preamble

The Governments of California, British Columbia, Oregon and Washington,

Pursuant to the *Memorandum to Establish the Pacific Coast Collaborative* of June 2008, as provided for in Article 6;

Affirming our shared vision of Pacific North America as a model of innovation that sustains our communities and creates jobs and new economic opportunities for our combined population of 53 million;

Recognizing that the Pacific Coast is a region bound together by a common geography, shared infrastructure and a regional economy with a combined GDP of US $2.8 trillion, which makes it the world's fifth largest;

Acknowledging the clear and convincing scientific evidence of climate change, ocean acidification and other impacts from increasing concentrations of carbon dioxide in the atmosphere, which threaten our people, our economy and our natural resources;

Emphasizing that states and provinces around the world are battling climate change through technology innovation and actions that limit greenhouse gas emissions and other air pollution while creating economic growth, consumer savings and new jobs;

Celebrating that our own governments have reduced greenhouse gas emissions by adopting regulatory, policy and market-based measures that shift energy generation to clean and renewable sources, manage energy use through greater efficiency and conservation, and enable and promote consumer choice for clean vehicles;

Supporting positive federal action to combat climate change, including President Obama's climate action plan and proposed rules to limit greenhouse gas emissions from power plants;

Joining the growing international convergence on the need to secure an international agreement to reduce global greenhouse gas emissions, including discussions at the coming Conference of Parties meetings in Warsaw (2013), Lima (2014) and Paris (2015); and

Agreeing that meaningful coordination and linkage between states and provinces across North America and the world on actions to reduce greenhouse gas emissions can improve the effectiveness of these actions, increase their overall positive impact and build momentum for broader international coordination to combat climate change.

> Our agreement, which established the Pacific Coast Action Plan on Climate and Energy, represents a region wide commitment to air quality, clean fuels, carbon pricing, and clean-energy jobs. But it also respects that we have different approaches to reaching our shared goals. California's carbon pricing program uses an economy wide cap-and-trade system, while British Columbia has a revenue-neutral carbon tax program. Oregon is building on existing programs to set a price on carbon emissions. Washington is developing a carbon market program, including consultations with stakeholders.

While the plan is voluntary and non-binding, it does communicate the strong sense of policy dedication to renewable energy technologies and to the reduction of greenhouse gases in this region.

2.8 California Energy Policies

In 1998, the Energy Resources Conservation and Development Commission (usually called the California Energy Commission), was placed in charge of a new Renewable Energy Program. The program was aimed at helping to increase total renewable electricity production in the state. It provided market-based incentives for new and existing utility-scale facilities powered by renewable energy, offered consumer rebates for installing new wind and solar renewable energy systems, and helped to educate the public regarding renewable energy.

The California Energy Commission, the Public Utilities Commission (CPUC), and the Consumer Power and Conservation Financing Authority (CPA), were responsible for proposing and approving the California Energy Action Plan. The first action plan was approved in 2003, and it established goals and actions to ensure that electrical power and natural gas supplies are achieved and provided through cost-effective and environmentally sound policies. Updates to the original plan were approved in 2005 and 2008, reflecting changes on the political scene (such as the Bill 32), studies, and evaluation reports.

California's Public Resources Code Section 25302 requires that every 2 years the Energy Commission submit to the Governor an integrated energy policy report that offers an overview of major energy issues and trends facing the state. The 2013 report focused, among other things, on renewable energy, in particular on impediments to procuring biomass in the state, and on possible solutions, as well as on bioenergy development. From its creation in 2006, the program funded "grid-connected, solar/photovoltaic electricity systems under 30 kilowatts on homes and businesses in the investor-owned utilities' service areas, wind systems up to 50 kW in size, fuel cells (using a renewable fuel), and solar thermal electric" (California Energy Commission 2015). In 2007, the California Renewable Energy Program became focused on supporting small wind systems and renewable fuel cells. In addition, the solar portion of the program ended and was replaced with the "Go Solar California" campaign.

The state of California, through its Renewable Portfolio Standard (RPS) Program, has been steadily increasing the percentage of renewable energy used in the state.

Table 2.3 Timeline of California's renewables portfolio standard

Year	Action
2002	Senate Bill 1078 establishes the RPS program, requiring 20% of retail sales from renewable energy by 2017
2003	Energy Action Plan I accelerated the 20% deadline to 2010
2005	Energy Action Plan II recommends a further goal of 33% by 2020.
2006	Senate Bill 107 codified the accelerated 20% by 2010 deadline into law.
2008	Governor Schwarzenegger issues Executive Order S-14-08 requiring 33% renewables by 2020
2009	Governor Schwarzenegger issues Executive Order S-21-09 directing the California Air Resources Board, under its AB 32 authority, to adopt regulations by July 31, 2010, consistent with the 33% renewable energy target established in Executive Order S-14-08
2011	Senate Bill X1–2, signed by Gov. Edmund G. Brown, Jr., codifies 33% by 2020 RPS

Source: California Energy Commission (2015)

When it was created in 2003, the program *recommended* that the state reach the goal of 20% renewable energy usage by 2010, a target that was codified by Senate Bill 107. Governor Schwarzenegger, the Energy Commission, and the California Public Utilities Commission endorsed this enhanced goal for the state as a whole (see Table 2.3; California Energy Commission 2015).

Achieving these renewable energy goals became even more important with the enactment of AB 32, the California Global Warming Solutions Act of 2006. Because of this Act, in 2006 the RPS calendar was accelerated. Senate Bill 107 *required* the target to be reached by 2010. In 2011, Governor Jerry Brown expanded the RPS by signing Senate Bill X1–2, which broadened the program's scope to include all utilities (including municipal utilities) and increased the renewables target to 33% by December 31, 2020. Intermediate targets were also set: retail sellers must use at least 20% renewable energy by 2013 and 25% renewable energy by 2016. Finally, Bill X1–2 determines that Tradable Renewable Energy Credits (TRECs) can be used for 25% of the requirement for 2013 and for 10% of the requirements for 2017. Moreover, state officials have debated a new target to be achieved by 2050, namely to reduce emissions by 80% (California Energy Commission 2015).

The "Go Solar California" campaign was created in 2007, with the goal of generating 3000 MW of solar energy statewide, with a $3.35 billion investment by the end of 2016; it also aimed at creating 585 million therms of solar hot water systems by end of 2017. The Go Solar California Campaign is composed of two programs: the Energy Commission's New Solar Homes Partnership and the California Public Utilities Commission's California Solar Initiative. The New Solar Homes Partnership offers $400 million dollars in incentives to encourage highly efficient solar installations in new residential areas. The California Solar Initiative, on the other hand, transferred the consumer rebate program from the California Energy Commission to the utility companies. The CSI has a budget of $2 billion, and it provides cash back to new and existing residential and commercial facilities for installing solar systems. The CSI-*Thermal* Program offers cash rebates of up to $2719 for solar water heating systems on single-family homes. Multifamily and Commercial properties qualify for rebates of up to $500,000 (California Energy Commission 2015).

In June 2010, Governor Jerry Brown proposed an action plan for investment in renewable energy technology and the creation of more than half a million green jobs. This program was based on the contention that clean energy and energy efficiency are central elements in the creation of hundreds of thousands of jobs; that they create more jobs than do gas, oil, and coal energy; and that they have continued to grow, even during economic downturns. Governor Brown's plan proposed the following:

- Building 12,000 megawatts of Localized Electricity Generation
- Building 8000 Megawatts of Large Scale Renewables and Necessary Transmission Lines
- Dealing with Peak Energy Needs and Develop Energy Storage
- A Timeline to Make New Homes and Commercial Buildings Zero Net Energy
- Making Existing Buildings More Efficient
- Adopting Stronger Appliance Efficiency Standards
- Developing More Cogeneration Projects to increase combined heat and power production by 6500 megawatts
- The Appointment of a Renewable Energy Jobs Czar (California Energy Commission 2015).

The California Clean Energy Jobs Act was adopted in December 2013. It makes changes to corporate income taxes and, except as specified, provides $550 million annually from the General Fund to the Clean Energy Job Creation Fund, or the Job Creation Fund, for five fiscal years, beginning with the 2013–14 fiscal year. Senate Bill 73 describes how the funds will be allocated to California's K-12 schools and community colleges. The California Energy Commission is working with the California Department of Education, California Workforce Investment Board, California Conservation Corps, and the Department of General Services in order to implement the California Clean Energy Jobs Act (California Energy Commission 2015).

In 2013 California adopted the nation's first energy storage mandate. The plan requires three of the largest utilities—PGE, Southern California Edison, and San Diego Gas and Electric—to expand their capacity to store electricity, including renewable energy generated from solar and wind. The state's three investor-owned utilities must collectively buy 1.3 gigawatts—or 1325 megawatts—of energy storage capacity by the end of 2020. The decision to mandate energy storage is expected to spur innovation in emerging storage technologies, from batteries to flywheels (California Energy Commission 2015).

Customers who install small solar, wind, biogas, and fuel cell generation facilities (1 MW or less) to serve all or a portion of onsite electricity needs are eligible for the state's net metering program (NEM). Over a 12-month period, customer have to pay only for the net amount of electricity used from the utility over-and-above the amount of electricity generated by their solar systems. NEM is an important element of the policy framework supporting direct customer investment in

Table 2.4 Timeline of Oregon's climate change efforts

Year	Action
1988	The Oregon Task Force on Global Warming recommends creating a permanent Global Warming Management Group, reduction of net greenhouse gases, and having state agencies building climate change into their plans and programs.
1992	Oregon adopts a benchmark to hold state's carbon dioxide emissions to 1990 levels.
1997	Oregon becomes first state to control emissions of carbon dioxide. The Energy Facility Siting Council is given authority to set carbon emissions standards for new energy facilities.
2003	Oregon Governor Kulongoski with California and Washington Governors establishes the West Coast Governors' Global Warming Initiative.
2004	An Advisory Group on Global Warming is appointed by the governor to produce the *Oregon Strategy for Greenhouse Gas Reductions*.
2006	Governor Kulongoski issues Executive Order on Sustainability for the twenty-first Century, calling for the Department of Administrative Services to coordinate an interagency team to inventory greenhouse gases.
2007	Oregon Legislature passes HB 3543 to reduce greenhouse gas emissions to 1990 levels by 2010; achieve a 10% reduction in greenhouse gas emissions by 2020; achieve "climate stabilization" emission levels by 2050, 75% below 1990 levels.
2009	Oregon Legislature passes SB 101 to prevent new conventional coal plants; HB 2186 to reduce carbon intense fuels; and HB 2078 to provide incentives for emission free vehicles.
2015	Governor Kate Brown signs SB 324, which continues a program begun in 2009, it creates a market for carbon credits in Oregon's transportation fuel industry, regulated by the state Department of Environmental Quality.

Source: State of Oregon Energy Plan 2011–2013 (2011)

grid-tied distributed renewable energy generation, including customer-sited solar PV systems. NEM is applicable to all sectors: commercial, industrial, residential, general public/consumer, nonprofit, schools, local government, state government, federal government, multi-family residential, low-income residential, agricultural, and institutional (California Energy Commission 2015).

The CPUC's Self-Generation Incentive Program (SGIP) provides incentives to support existing, new, and emerging distributed energy resources. The SGIP provides rebates for qualifying distributed energy systems installed on the customer's side of the utility meter. Qualifying technologies include wind turbines, waste heat to power technologies, pressure reduction turbines, internal combustion engines, microturbines, gas turbines, fuel cells, and advanced energy storage systems. The SGIP was initially conceived of as a peak-load reduction program in response to the energy crisis of 2001. Assembly Bill 970 designed the Program as a complement to the California Energy Commissions' Emerging Renewables Program, which focused on smaller systems than the SGIP. Since 2001, the SGIP has evolved significantly. It no longer supports solar photovoltaic technologies, which were moved under the purview of the California Solar Initiative after its launch in 2006. SGIP has also been modified to include energy storage technologies, to support larger projects, and to provide an additional 20% bonus for California-supplied products.

Senate Bill 412 (2009) modified the focus on the Program to include greenhouse gas reductions (California Energy Commission 2015).

2.9 Oregon Renewable Energy Policies

Oregon's climate and renewable energy policy agenda has been very active over the years (see Table 2.4). Beginning in 1988, the Oregon Task Force on Global Warming, established by the governor, recommended reductions in greenhouse gas emissions and required all state agencies to incorporate climate change scenarios into their programs and plans. In 1992, the legislature adopted a benchmark to keep carbon dioxide emissions at 1990 levels or lower, and in 1997, Oregon was the first state to control emissions of carbon dioxide.

With the election of Governor Kulongoski in 2003, Oregon sought to become a leader in renewable energy technology and use. Governor Kulongoski signed the nonbinding West Coast Governors' Global Warming Initiative in 2003 and in 2004 appointed an Advisory Group on Global Warming, with a broad spectrum of stakeholders participating. This group then produced the *Oregon Strategy for Greenhouse Gas Reductions*, which was used to guide renewable energy policy.

The Oregon Renewable Portfolio Standard was enacted in 2007 through Senate Bill 838 and requires Oregon utilities to deliver a percentage of their electricity from renewable resources by 2025. Eligible resources include biomass, geothermal, hydropower, ocean thermal, solar, tidal, wave, wind, and hydrogen (if produced from any of these sources). The RPS is 25% reduction by 2025 for utilities with 3% or more of the state's load, 10% reduction by 2015 for utilities with 1.5–3% of the state's load, and 5% reduction by 2025 for utilities with less than 1.5% of the state's load. Oregon utilities may also comply with the RPS by using a combination of the following options: (1) building an eligible facility and retaining renewable energy certificates (REC) output from these facilities; (2) buying power and REC from another eligible facility; (3) buying "unbundled" RECs separate from power (some limits on this); and (4) making "alternative compliance payments" with options to use these funds for construction of an eligible facility in the future.

In 2009, Oregon passed SB 101 to prevent the construction of new conventional coal plants, HB 2078 to provide incentives for emission free vehicles, and HB 2186 to reduce carbon-intense fuels. With the election of Governor Kitzhaber in 2011, Oregon was poised to continue an aggressive approach to renewable energy development and climate change mitigation efforts. However, scandal related to the governor's partner and her role in renewable energy policy led to his resignation in 2015. The new Governor, Kate Brown, is also committed to continuing Oregon's leadership in renewable energy development. Governor Brown signed SB 324 in 2015, which continues a program begun in 2009 to create a market for carbon credits in Oregon's transportation fuel industry, regulated by the state Department of Environmental Quality.

Other notable Oregon renewable energy policies include the somewhat controversial Oregon Business Tax Credit program, which was replaced by the Energy Incentive Program in 2014. The original tax credit program provided up to 50% of eligible project costs over 5 years, or 35% over 10 years. The program also allowed unused credits to carry forward for 8 years, and allowed businesses to sell their tax credits for lump-sum cash payments. Many of Oregon's largest wind farms received the tax credit, along with projects related to conservation, renewables, homes, buildings, transportation, weatherization, and recycling. Because of perceived abuses with the program, the tax credit program was replaced by a new Energy Incentive Program in 2014. According to an investigation by *The Oregonian* newspaper (Sickinger 2014): "Loose rules and lax administration of the program meant that credits were issued to projects that quickly went bankrupt or never operated, to projects that would have been built without the credits, and to developers who gamed the state's loose rules to qualify for multiple credits." The new Energy Incentive Program also offers tax credits, but is far less ambitious and has much more oversight.

Other policies and programs include (1) a biomass producer or collector tax credit to support the production, collection and transportation of biomass for energy production; (2) a net metering program that allows utility customers to generate their own electricity by installing photovoltaic or wind energy systems; (3) tax credits for renewable energy equipment manufacturers; (4) energy conservation tax credits for new projects; residential tax credits for making houses more energy efficient; (5) a state energy loan program for 5- to 20-year loans at $20,000 to $20 million dollars; and (6) a variety of programs to incentivize small wind projects, rural energy projects, etc.

2.10 Washington Renewable Energy Policies

While non-hydroelectric renewables are not a primary source of electricity in Washington, the state has received high rankings by nongovernmental organizations (NGOs) and media organizations for promoting their development and adoption, as have California and Oregon. One example is *Greenopia* (2011), an online consumer's directory for "green, sustainable and socially conscious, daily purchase decisions." Employing data from a number of federal and state government agencies and other environmental NGOs, *Greenopia* compiled an index of sustainability in 2011 based on air quality, water quality, recycling rate, number of green businesses, LEED buildings, and per-capita rates for greenhouse gases emissions, energy use and water consumption, and waste generation. Washington, along with California and Oregon, were in the top ten "greenest" states.

Additionally, *Forbes Magazine* has ranked states in terms of "greenness" using six equally weighted indicators that included air quality, carbon footprint, hazardous waste management, energy consumption, policy initiatives, and water quality (*Forbes* 2007). According to *Forbes,* the top-ranked states include both Oregon and

Table 2.5 Brief timeline of Washington's renewable energy efforts

Year	Action
2000	RCW 19.29A.050 required that all retail electric suppliers in the state disclose to the public the fuel mix details of their electricity generation on an annual basis. Larger utilities are required to release this information twice a year.
2001	EHB 2247 required that utilities allow customers the ability to purchase renewable energy if they are serving more than 25,000 customers. Solar, wind, solar, landfill gas, geothermal, wave or tidal action, biomass, and fish-friendly hydropower are among eligible renewables.
2005	RCW 19.260.010 created minimum efficiency standards for 12 products. In May 2009, HB 1004 added several more products to the list. These products include household water heaters, bottle-type water dispensers, pool heaters, hot water dispensers, residential pool pumps, etc. These standards are only for goods within the state and not for products manufactured in and sold outside of Washington.
2006	Energy Independence Act (19.285 RCW) established an energy efficiency resource standard. The act requires the increased use of renewable energy and conservation through new requirements for utilities. It also requires utilities serving more than 25,000 customers to participate in all cost-effective energy conservation.
2006	Renewable Portfolio Standard established to increase the amount of renewable energy production. Washington met its interim RPS goals of 3% in 2012, 9% by 2016, and is on track to meets its 2020 goal of 15% renewable electricity generation.
2007	Governor Gregoire over-ruled a no-vote from Kittitas County to install 65 wind turbines northwest of Ellensburg, Washington leading to much political animosity among residents.
2014	Governor Inslee issues Executive Order 14–04 establishes Carbon Emissions Reduction Taskforce to provide recommendations on the design and implementation of carbon limitations and markets; also directs the Department of Transportation, Department of Commerce and Department of Enterprise Services to collaborate and coordinate with other agencies to address greenhouse gas emissions.

Source: Washington State Energy Office (2015a, b)

Washington because of "low carbon dioxide emissions per capita, strong policies to promote energy efficiency and air quality" as well as having the highest number of LEED-certified buildings per capita (*Forbes* 2007, 1).

Table 2.5 provides a brief timeline for some major renewable energy efforts and policy in Washington. In 2000, the legislature passed a bill that required all retail electric suppliers to disclose the fuel source of their electricity, with the largest utilities required to disclose twice a year. In 2001, the legislature passed additional legislation requiring utilities to offer renewable energy options to customers in larger markets (25,000+ customers).

During 2006 the Washington legislature passed the Energy Independence Act, which required the increased use of renewable energy sources and conservation efforts by utilities. In addition, the state adopted a renewable portfolio standard. Washington was the second state to pass a renewable energy standard by the initiative process. Initiative 937 requires that certain utility companies get 15% of their electricity from renewable resources by 2020; in addition, the companies must identify and embrace cost-effective energy conservation strategies, as well as review and update their effectiveness every 2 years. Renewable energy is defined as coming from water, wind, sun, geothermal sources, landfill gas, biodiesel, and biomass. Eighty-four percent of Washington's 62 utility companies are required to comply. All utilities are also required to meet intermediate targets of at least 3% renewable energy by 2012 and 9% by 2016.

As the case study in Chap. 4 describes, a significant event took place in 2007 concerning renewable energy development in Washington. Then-Governor Gregoire overruled a negative vote by county commissioners in Kittitas County in regard to a 65-turbine wind farm. While Washington allows counties to regulate wind farms, the state's Energy Facility Site Evaluation Council can override local opposition, which it did in this case, with the governor's strong support. Part of the driving force behind the council's and the governor's decision was Initiative 937 requiring larger utilities to increase their renewable resources. This decision has continued to reverberate during the siting of other wind farms in the state, in that local communities fear that the state will override their preferences when politically feasible (Durlin 2009).

In 2014, Governor Inslee issued Executive Order 14–04 that established a Carbon Emissions Reduction Taskforce (CERT) to provide recommendations on the design and implementation of carbon limitations and markets. It also directed the Department of Transportation, Department of Commerce, and Department of Enterprise Services to collaborate and coordinate with other agencies to address greenhouse gas emissions. On November 14, 2014, CERT submitted a final report to the Governor with some guidelines, caveats, and suggestions for creating a carbon market in Washington. The Executive Summary of the report concluded with the following passage (CERT 2014: 3):

> Overall, a policy design going forward needs to address an integrated approach which considers such items as: land-use policies; equitable transit oriented development; alternatives to current single occupancy vehicles such as adequate transit, zero emissions vehicles, and alternative fuel infrastructure; the different needs of rural communities and industries that require long distance travel; the effects on and needs of low-income communities; and the incentives for the electric sector to take on activities like vehicle electrification infrastructure that reduce societal emissions while potentially increasing emissions in the electric sector

Similar to California and Oregon, there are other programs and policies in place in Washington to encourage the development of renewable energy resources and climate change mitigation. These programs and policies include, but are not limited to the following: a 75% exemption from tax for the sales of equipment used to generate electricity using fuel cells, wind, sun, biomass energy, tidal or wave energy, geothermal, anaerobic digestion or landfill gas; a reduced business and occupation tax rate for Washington manufacturers of solar-electric modules or silicon components of those systems; an exemption from state sales and use taxes for waste vegetable oil, specifically, cooking oil gathered from restaurants or commercial food processors for producing biodiesel for personal use; at least 2% of all diesel fuel sold in Washington must be biodiesel or renewable diesel; and, at least 2% of the total gasoline sold in the state must contain ethanol.

2.11 Idaho Renewable Energy Policies

Idaho does not have an official target for renewable energy use, but with Executive Order 2007–2, Governor C.L. "Butch" Otter established the 25X25 Renewable Energy Council, which was assigned to the Idaho Office of Energy Resources. The state does offer some renewable energy incentive programs in conjunction with utilities. For example, Idaho does not have a statewide net-metering policy. However, the state's three investor-owned utilities—Avista Utilities, Idaho Power and Rocky Mountain Power—have each developed net-metering programs that have been approved by the Idaho Public Utilities Commission. Each of the utilities offers net metering to customers that generate electricity using solar, wind, hydropower, biomass, or fuel cells, with residential systems limited to 25 kilowatts. The utilities also have a limit on aggregate net-metered capacity to 0.1% of the utility's peak demand in a baseline year, and they restrict any single customer from generating more than 20% of the aggregate capacity of all net-metered systems (American Council on Renewable Energy 2014).

Idaho also offers tax incentives in the form of an income tax deduction for the cost of residential renewable energy systems, as well as property tax exemptions. Idaho also has a low-interest loan program and a bond program. Idaho residents with homes built or under construction before 2002, or who had a building permit issued before January 1, 2002, qualify for an income tax deduction for 100% of the cost of installing new insulation or other approved energy efficiency improvements in an existing residence. Other energy efficiency measures that may be deducted are windows that are more energy efficient than existing windows, weather stripping and caulking, and duct sealing and insulation (American Council on Renewable Energy 2014).

In 2007, Idaho enacted a bill that restructured the method of taxation for producers of wind energy from a property tax to a tax on production. In 2008, Idaho enacted further legislation to include geothermal energy producers under this method of taxation. Under these policies, commercial wind operators and geothermal energy producers, excluding those regulated by the Idaho Public Utilities Commission, are exempt from paying taxes on real estate, fixtures, or property related to their renewable energy systems. In lieu of property taxes, however, wind and geothermal energy producers must instead pay a tax of 3% of gross energy earnings (American Council on Renewable Energy 2014).

2.12 Conclusion

Encouraging the development of renewable energy is a primary focus of energy policy throughout the United States and for California, Oregon, and Washington particularly. As this chapter has shown, all three jurisdictions have actively pursued a policy agenda that promotes wind, solar, and biomass energy. Combined with direct efforts to expand renewable energy, each of the states has adopted aggressive Renewable Portfolio Standards and promoted programs and policies that target a corresponding reduction in fossil fuels. Each of the states has also engaged in regional efforts with British Columbia to promote the Pacific Coast Action Plan on Climate and Energy, further demonstrating the region's commitment to clean energy solutions. Although the arguably more conservative Idaho has done less to promote such policies, it offers some renewable energy incentive programs in conjunction with utilities. Given this favorable climate for renewable energy in the study region, how successful has the siting and development of new AET projects been and why—or why not?

Recent research investigating public preferences for energy sources indicates an overwhelming support for wind and solar (Steel et al. 2015). However, community responses to locally proposed renewable energy projects periodically conflict with policy goals and the broader public sentiment. Several case studies detail controversial wind siting proposals that have elicited vocal opposition from local individuals and organizations (e.g., Devine-Wright 2012, 2005). In some communities, pockets of NIMBY (Not in My Backyard) resistance developed into organized opposition that contributed to project failure. Nevertheless, other studies have highlighted communities that expressed strong levels of acceptance, with some even proactively courting wind projects, in a phenomenon sometimes referred to as YIMBY, or "Yes in My Backyard" (Warren and McFadyen 2010).

While local opposition to renewable energy development is sometimes ascribed to selfish parochialism and labeled as a NIMBY response, research has revealed a complex range of factors influencing community acceptance of locally proposed wind facilities. In an exhaustive review of the literature, Petrova (2013) identified four key areas that affect community acceptance of wind project proposals: (1) visual and landscape concerns, (2) socioeconomic concerns, (3) environmental concerns, and (4) procedural factors. However, other recent research also has shown that actual experience with wind power installations (as opposed to hypothetical installations) does not elicit the same level of negativity (Krause et al. 2016). Given the policy priorities of California, Oregon, and Washington, deeper understanding of which factors and processes most strongly propel support or opposition to wind facility siting is of both practical and conceptual importance, and will have relevance to other types of AETs as well.

The remainder of Part II provides four case study chapters, each consisting of an in-depth analysis concerning the siting of a different type of renewable energy technology in three of the study states. The first two case studies concern highly controversial projects where there were environmental, social, and economic tradeoffs. Chapter 3, a case study of the failed Calico Solar Project, illustrates a successful challenge to a grid-scale solar energy park in the California desert. Chapter 4 presents a study of the successfully sited Kittitas Valley Wind Project in Washington—albeit a case wherein the state government overrode local concerns about the aesthetics of and potential environmental damage from a large wind farm. The third and fourth case studies examine nascent renewable technologies currently being developed in Oregon, including proposed wave energy projects off Oregon's coast

(Chap. 5), and a geothermal energy installation near Oregon's Newberry Volcanic Crater (Chap. 6). All four projects involve examples of environmental versus renewable energy "tradeoffs," as well as procedural concerns in the siting process.

References

American Council on Renewable Energy. (2014). *Renewable energy in the 50 states: Western region*. http://www.acore.org/publications-category/3868-renewable-energy-in-the-50-states-western-region. Accessed 15 Aug 2016.

Brown Jr., E. G., Clark, C., Inslee, J. & Kitzhaber, J. (2014, December 11). A Pacific coast model for addressing climate changes. *Los Angeles Times*. http://www.latimes.com/opinion/op-ed/la-oe-1212-brown-west-coast-climate-change-plan-20141212-story.html. Accessed 15 Aug 2016.

California Energy Commission. (2015). *California renewable energy overview and programs*. http://www.energy.ca.gov/renewables/. Accessed 15 Aug 2016.

California Environmental Protection Agency. (2014). *California reformulated gasoline*. Sacramento: California Air Resources Board.

CERT. (2014, November 14). Carbon emissions reduction taskforce: Report to the Washington State Governor's Office. http://www.governor.wa.gov/sites/default/files/documents/CERT_Final_Report.pdf. Accessed 15 Aug 2016.

Devine-Wright, P. (2005). Local aspects of renewable energy development in the UK: Public beliefs and policy implications. *Local Environment, 10*(1), 57–69.

Devine-Wright, P. (2012). Explaining "NIMBY" objections to a power line: The role of personal, place attachment and project-related factors. *Environment and Behavior, 45*(6), 761–781.

Durlin, M. (2009, February 18). Wind setbacks: Local governments grapple with where to put wind farms. *High Country News*. https://www.hcn.org/issues/41.3/wind-setbacks. Accessed 15 Aug 2016.

Forbes Magazine. (2007). *America's greenest states*. http://www.forbes.com/2007/10/16/environment-energy-vermont-biz-beltway-cx_bw_mm_1017greenstates.html. Accessed 15 Aug 2016.

Krause, R., Pierce, J., & Steel, B. S. (2016). The impact of auditory and visual experience with wind turbines on support for wind production and proximity-based opposition. *Society and Natural Resources*. doi: 10.1080/08941920.2016.1171936.

Petrova, M. A. (2013). NIMBYism revisited: Public acceptance of wind energy in the United States. *WIREs Climate Change, 4*, 575–601. doi:10.1002/wcc.250.

Sickinger, T. (2014, July 2). Oregon's business energy tax credit is officially dead, but its liability lives on. *The Oregonian*. http://www.oregonlive.com/business/index.ssf/2014/07/oregons_business_energy_tax_cr.html. Accessed 15 Aug 2016.

State of Oregon Energy Plan 2011–13. (2011). Oregon Department of Energy. http://www.oregon.gov/energy/docs/reports/legislature/2011/energy_plan_2011-13.pdf. Accessed 15 Aug 2016.

Steel, B. S., Pierce, J. C., Warner, R. L., & Lovrich, N. P. (2015). Environmental value considerations in public attitudes about alternative energy development in Oregon and Washington. *Environmental Management, 55*, 634–645.

U.S. Department of Energy. (2012a). *State electricity profiles 2010*. Washington, DC: U.S. Energy Information Administration. http://www.eia.gov/state/data.cfm?sid=ID#ConsumptionExpenditures. Accessed 15 Aug 2016.

U.S. Department of Energy. (2012b). *California state profiles and energy estimates*. Washington, DC: U.S. Energy Information Administration. http://www.eia.gov/state/?sid=CA#tabs-1. Accessed 15 Aug 2016.

U.S. Department of Energy. (2012c). *Table F30: Total energy consumption, price, and expenditure estimates*. Washington, DC: U.S. Energy Information Administration. http://www.eia.gov/state/seds/data.cfm?incfile=/state/seds/sep_fuel/html/fuel_te.html&sid=US&sid=CA. Accessed 15 Aug 2016.

U.S. Department of Energy. (2012d). *State electricity profiles 2010*. Washington, DC: U.S. Energy Information Administration. http://www.eia.gov/state/rankings/?sid=US#/series/101. Accessed 15 Aug 2016.

U.S. Department of Energy. (2012e). *State electricity analysis*. Washington, DC: U.S. Energy Information Administration. http://www.eia.gov/state/analysis.cfm?sid=ID. Accessed 15 Aug 2016.

U.S. Department of Energy. (2012f). *Oregon state profiles and energy estimates*. Washington, DC: U.S. Energy Information Administration. http://www.eia.gov/state/?sid=OR#tabs-5. Accessed 15 Aug 2016.

U.S. Department of Energy. (2012g). *State electricity analysis*. Washington, DC: U.S. Energy Information Administration. http://www.eia.gov/state/analysis.cfm?sid=WA. Accessed 15 Aug 2016.

U.S. Department of Energy. (2014). *U.S. states: State profiles and energy estimates*. Washington, DC: U.S. Energy Information Administration. http://www.eia.gov/state/?sid=US. Accessed 20 May 2015.

Warren, C. R., & McFadyen, M. (2010). Does community ownership affect public attitudes to wind energy? A case study from south-west Scotland. *Land Use Policy, 27*(2), 204–213.

Washington State Energy Office. (2015a). *Energy efficiency programs overview*. http://www.commerce.wa.gov/Documents/Commerce-Energy-Efficiency-Overview-2014.pdf. Accessed 15 Aug 2016.

Washington State Energy Office. (2015b). *A legacy of clean energy, an opportunity for climate action*. http://www.commerce.wa.gov/Documents/gov-inslee-legacy-clean-energy--opprtunity-2013.pdf. Accessed 15 Aug 2016.

Chapter 3
The Calico Solar Project

Abstract The large proposed Calico Solar Project generated environmental concerns and a lawsuit by environmental groups (i.e., Defenders of Wildlife, the Natural Resources Defense Council, and the Sierra Club), and was ultimately terminated by the developer K Road Power. The failed project, located in California's Mohave Desert, was either an example of a successful campaign by environmental groups to stop development in an environmentally sensitive area with threatened wildlife (e.g., Desert Tortoise), or the result of changing market conditions where the project became economically unviable for the developer. Regardless of the ultimate reason for project failure, the case study provides a classic example of environmental concerns versus renewable energy development.

Keywords Solar energy • Environmental groups • Threatened wildlife • Market conditions

3.1 Introduction

> *Green technology faces a number of barriers -- high capital costs, entrenched incumbents, skepticism, bureaucratic and political roadblocks. But, please, does the environmental community have to continue to be a problem? The latest example comes in a lawsuit filed by the Sierra Club to delay an already-faltering solar plant slated for Calico, California.* (From "Dear Environmental Community: Please Shut Up" (2011), Michael Kanellos)

California plays a central role in the U.S. solar industry, thanks in part to the state's climate and geographical factors, as well as policy requiring that 33% of the state's electricity come from renewable sources by 2020 (McBride 2011; Woody 2010; Sahagun 2012). When Kenneth Salazar became U.S. Secretary of the Interior in 2009, several large solar projects proposed on public lands were waiting for approval, among them the Calico Solar Energy Project in San Bernardino County, California. The federal government set a goal of speeding up approvals for some of these projects, with an objective of having several projects approved before the end of 2010, so that they could benefit from the Federal Loan Guarantee Program. In addition, President Barack Obama had also signed the American Recovery and Reinvestment Act (ARRA) in 2009. The ARRA offered additional incentives for solar projects that could begin construction before the end of 2010 (Reid 2011).

J.C. Pierce, B.S. Steel, *Prospects for Alternative Energy Development in the U.S. West*, Environmental Challenges and Solutions 4,
DOI 10.1007/978-3-319-53414-5_3

Table 3.1 California approved projects

Company	Project name	Capacity (MW)	Technology	Location (county)
Genesis Solar LLC	Genesis Solar Project	250	Parabolic trough	Riverside
Palo Verde Solar	Blythe Solar Project	1000	Parabolic trough	Riverside
Tessera Solar	Calico Solar	664	Solar dish (SunCatcher)	San Bernardino
Tessera Solar	Imperial Valley Solar Project	709	Solar dish (SunCatcher)	Imperial
BrightSource Energy Inc.	Ivanpah Electric Generating System	370	Power tower	San Bernardino
Chevron Energy Solutions	Chevron Lucerne Valley Solar Project	45	Photovoltaics	San Bernardino

Even though there was pressure to "fast-track" approvals, project developers and the Department of the Interior (DOI) still had to conduct the environmental review process. As Reid (2011) explained, "most of the US public lands are managed for conservation purposes, they are strictly governed by environmental regulations and it's difficult to get anything approved for construction, even a solar field. In order to gain approval, projects had to get a yes vote from a variety of environmental groups, such as the Sierra Club" (Reid 2011: 80). Thus, in order to see their projects approved, solar developers had to work closely with environmental groups. As noted in a report by the Defenders of Wildlife, Sierra Club, and NRDC (Natural Resources Defense Council), "Seven utility-scale solar projects totaling a little more than 23,000 acres of public lands have been permitted in the California desert since 2009. Together, these facilities could generate 2924 megawatts of clean, renewable electricity" (Defenders 2011a: 1).

Working with the environmental groups, the developers have reduced the potential negative impacts to wildlife for seven species, "resulting in the permitting of some 2595 megawatts on public lands in California" (Defenders 2011b: 3). From 2009 to the present, in collaboration with environmentalists, the California Energy Commission has permitted six out of the seven larger-scale solar plant projects fast tracked by DOI. Organizations such as Natural Resources Defense Council (NRDC), Sierra Club, and Defenders of Wildlife worked with project developers to reduce the potential impacts of some of these projects. Collaboration of NRDC and other agencies helped the permitting of almost 90% of the energy allowed in the California desert (McBride 2011; Sahagun 2012).

These six approved projects were picked by DOI for a "fast-track" approval process because they had completed several environmental impact statement (EIS) drafts with environmental groups (Table 3.1). The fifth of these projects to be approved was the Calico Solar Project, which qualified for $1.75 billion through the Department of Energy (DOE) Loan Guarantee Program (Reid 2011). Interior Secretary Salazar, as quoted in Reid (2011: 81) noted, "These projects show what

great strides we are making' …[and] reflect [President] Obama's focus a
ment to supporting America's clean energy economy."

3.2 Case Study Project

As will be discussed below, the only project that was approved in this wave but encountered problems with environmentalists was the Calico Solar Project. As one observer characterized the project: "State and federal government officials have generally worked alongside environmental groups to help the budding solar industry build plants across the California desert. The fight over Calico was a highly visible exception" (Sahagun 2012). Environmentalists' efforts to work with proponents of the Calico Solar Project were not successful: "The conservation community met with the project developers no less than 10 times over three years to urge developers to move the project to less-sensitive locations identified by the groups and the Bureau of Land Management (BLM), but the developers have refused to relocate the project" (Defenders 2012a: 1). As Annie Notthoff (2012), a NRDC staff member, stated: "the Calico project is the **only** utility scale solar project on public or private land that NRDC and two of our key partners, Defenders of Wildlife and the Sierra Club, have challenged in court."

Although the NRDC wants a transition away from fossil fuel consumption, they consider that in order to be clean and green, solar power must be designed in a way that minimizes harms to natural and cultural resources. The NRDC main demand is that utility-scale renewable energy projects "start smart" by choosing locations that "have high clean-energy potential; minimal conflicts with wildlife, wild lands, water, and other important resources and uses of the surrounding environments (such as disturbed lands); and, wherever feasible, access to existing transmission" (Defenders 2011a, p. 4).

3.3 The Initial Application for the Calico Project

On December 2, 2008, Stirling Energy Systems submitted an Application for Certification to construct and operate the Stirling Energy Systems Solar One Project (SES Solar One) in San Bernardino County, California (California Energy Commission n.d.). On January 2010, the project formally changed its name to the Calico Solar Project. The applicant, SES Solar Three, was merged into SES Solar Six, and that surviving entity was re-named Calico Solar, which was a subsidiary of Tessera Solar (California Energy Commission n.d.). The proposed Solar One project would be constructed on an estimated 8230-acre site. According to the Bureau of Land Management (BLM) (2010a, p. ES-1):

Fig. 3.1 A picture of the proposed "Stirling" PV installation. This technology involves converting solar power to rotary power, which then drives "an electrical generator to produce grid quality electricity" (California Energy Commission 2013)

> The Calico Solar Project [was] an electric-generating facility with a [initially proposed] nominal capacity of 850 megawatts (MW) using concentrated solar power. The main objective of the Calico Solar Project [was] to provide clean, renewable, solar-powered electricity to the State of California. The electricity from the Calico Solar Project would assist the State in meeting its objectives as mandated by the California Renewable Portfolio Standard (RPS) Program and the California Global Warming Solutions Act. The project site [was] in the Mojave Desert in San Bernardino County, California, north of Interstate 40 (I-40), approximately 37 miles east of Barstow; approximately 57 miles northeast of Victorville; and approximately 115 miles east of Los Angeles.

The primary equipment for the generating facility would be solar dish Stirling systems or "SunCatchers" (Fig. 3.1). According to the California Energy Commission, the proposed technology would have included photo voltaic (PV) modules mounted on various horizontal-tracking systems to most effectively follow the sun (California Energy Commission 2013). Construction of the proposed 850-megawatt Stirling engine project was planned to begin in late 2010, if approved by the California Energy Commission (CEC) (California Energy Commission n.d.).

The organizations responsible for the permitting process, the BLM and the California Energy Commission, had signed a memorandum in 2007 that allowed them to jointly conduct environmental reviews of solar thermal power projects on BLM land in California. This review would be in compliance with the National Environment Policy Act, California Environmental Quality Act, and other federal and state laws and regulations pertaining to power generation sites (Bureau of Land Management 2010a). The agencies' roles are detailed in Table 3.2 (Bureau of Land Management 2010a, ES-2):

Table 3.2 Agency roles and responsibilities relevant to the Calico Project

Agency	Roles and responsibilities
U.S. Bureau of Land Management (BLM)	The BLM must comply with the Federal Land Policy and Management Act (FLMPA) of 1976, Section 211 of the Energy Policy Act of 2005 (EPAct), and the BLM's Solar Energy Development Policy. The EPAct requires that the Secretary of the Interior should seek to have approved a minimum of 10,000 MW of renewable energy generating capacity on public lands by 2015.
	BLM authority extends to BLM lands in the California Desert District, which are governed by the California Desert Conservation Area (CDCA) Plan. Because the CDCA would need to be amended to allow the Calico Solar Project on the project site, the BLM would also oversee the CDCA Plan amendment process.
California Energy Commission (CEC)	The CEC has the exclusive authority to certify the construction, modification, and operation of thermal electric power plants in California that generate 50 or more MW. The CEC certification is in lieu of any permit required by state, regional, or local agencies and by federal agencies to the extent permitted by federal law.
	The CEC must review power plant Applications for Certification (AFCs) to assess potential environmental impacts and compliance with applicable laws, ordinances, regulations, and standards.

In March 2010, the BLM and CEC released a joint Draft of Environmental Impact Statement proposing an amendment to the California Desert Conservation Area and the construction and operation of Calico Project (Bureau of Land Management 2010a). The amendment to the *California Desert Conservation Area (CDCA) Plan* was to allow solar energy development on a project site east of Barstow in California proposed by Calico Solar developers (Defenders 2011a). After the release of the initial assessment, the two organizations started working on different environmental review documents: BLM prepared a Final Environmental Impact Statement (FEIS) and the CEC prepared a Supplemental Staff Assessment (SSA).

3.4 Final Environmental Impact Statement (FEIS)

The BLM concluded the FEIS on August 2010. In the FEIS, the agency assessed impacts for six different alternative actions, including a proposed "no action" alternative. The impacts are listed in Table 3.3 (Bureau of Land Management 2010a, ES-11-20).

The BLM proposed a preferred alternative plan of action: "The Agency Preferred Alternative is a 6215-acre solar energy power plant project that was developed in the FEIS as a modification of the 8230-acre Proposed Action" (Bureau of Land Management 2010a, ES-3). The BLM-favored alternative was also the Environmentally Preferred Alternative, which would accommodate 34,000 SunCatchers generating 850 MW. The north boundary of the proposed project was adjusted to protect 1770 acres of habitat for wildlife (i.e., desert tortoises and bighorn sheep), as well as some rare plant species.

Table 3.3 Final environmental impact statement

Air quality and climate: Cumulative adverse short-term construction and operation impacts on air quality **Biological resources** Short-term and long-term direct and indirect adverse impacts on vegetation onsite due to construction and maintenance activities; Potential short-term and long-term direct and indirect adverse impacts from the spread of invasive, non-native and/or noxious weeds; Short-term and long-term direct and indirect adverse impacts on wildlife on the project site and in the immediate project vicinity due to increased trampling, predation, noise, light, traffic and habitat loss; Short-term and long-term direct and indirect adverse impacts on birds due to habitat loss, increased noise, lighting, glare, bird collisions, and electrocution; Short- and long-term direct and indirect adverse impacts on wildlife movement corridors and habitat linkages that are currently available on the project site because of the substantial barrier to wildlife movement that would be posed by the fenced solar field; Direct and indirect adverse short-term and long-term impacts to small-flowered androstephium due to reduction, fragmentation, and degradation of suitable habitats on the project site and in the immediate project vicinity; Direct and indirect adverse short-term and long-term impacts on banded Gila monsters, if they do occur on the project site; Direct and indirect short-term and long-term adverse impacts on desert tortoises on the project site, in the immediate project vicinity, and at translocation receptor sites, and to desert tortoise critical habitat within the Ord-Rodman DWMA; Direct and indirect, short-term and long-term adverse impacts on Mojave fringe-toed lizards and their habitat, on Bendire's thrashers, on burrowing owls, on golden eagles, on Le Conte's thrashers, and on mountain plovers; Negligible direct and indirect short-term and long-term adverse impacts on any Swainson's hawks occurring in the project vicinity;	**Climate change** Minor contributions to GHG emissions and reduction of soil carbon sequestration, but overall long-term, beneficial direct and indirect impacts due to a net reduction in GHG emissions across the electricity system; **Cultural resources** Significant adverse effects to three cultural resources with the potential for listing on the National Register of Historic Places; permanent long-term adverse direct, indirect, and cumulative impacts on other cultural resources due to construction and decommissioning activities and increased human access **Fire and Fuels** Significant adverse effects to three cultural resources with the potential for listing on the National Register of Historic Places; permanent long-term adverse direct, indirect, and cumulative impacts on other cultural resources due to construction and decommissioning activities and increased human access **Geology, Soils, Topography, and Mineral Resources** Long-term and short-term adverse direct and indirect impacts on soils from clearing of vegetation, diminished soil productivity from topsoil loss, loss of cryptobiotic soil and desert pavement, erosion, and compaction; incremental contribution to cumulative adverse impacts on soil resources Geologic hazards include the potential for ground shaking and a low likelihood of volcanic eruptions **Grazing and Wild Horses and Burros** Negligible direct, indirect, or cumulative impacts due to the low quality of grazing vegetation present and the fact that grazing is not currently occurring at the site Negligible direct, indirect, or cumulative impacts due to absence of designated HAs or HMAs, or any observations of wild horses and burros **Land Use** Direct and indirect long-term, adverse impacts due to the exclusion of other public land uses and the disturbance of 1180 acres of donated and acquired lands; incremental contribution to cumulative adverse impacts **Noise and Vibration** Direct and indirect adverse impacts created by short-term construction activities and by normal long-term operation of the solar power plant; no cumulative impacts due to location of other potential projects in the region immediately surrounding the sensitive receivers for the project

(continued)

Table 3.3 (continued)

Direct and indirect, short-term and long-term adverse impacts on American badgers, on desert kit foxes, and on Nelson's bighorn sheep on the project site and in the immediate project vicinity; Negligible short-term and long-term adverse impacts on special-status bats that forage over the project site; **Recreation** Direct and indirect short-term and long-term adverse impacts due to loss of recreational access to the project site, reducing scenic values and altering the recreational experience; incremental contribution to cumulative adverse impacts **Socioeconomics and Environmental Justice** Negligible beneficial short-term and long-term direct and indirect impacts from increased employment and potential increase in local population; Negligible direct, indirect, and cumulative impacts on school facilities since enrollment in local school districts is not anticipated to increase **Special Designations** Short-term and long-term indirect impacts on the wilderness values select WAs and WSAs by changing the natural and undisturbed landscape; Short-term and long-term indirect impacts on the Pisgah Crater ACEC because of the closure of existing OHV routes on the project site, reduced access to open space and potential translocation of desert tortoise into this area; Direct and indirect impacts on the Ord Rodman DWMA because of the potential translocation of desert tortoise into this area; Short-term and long-term adverse direct and indirect impacts on resources for which the land was acquired or accepted by donation; incremental contribution to adverse cumulative impacts;	**Public Health and Safety and Hazardous Materials** Minimal potential for on-site and off-site direct and indirect impacts due to handling and storage of hazardous materials, including hydrogen; no cumulative impacts due to small amounts and low hazard of the hazardous chemicals to be stored at the facility Direct and indirect impacts due to increase in disposal of non-hazardous wastes; no cumulative impacts due to modest quantities of waste and employment of waste recycling **Traffic and Transportation** Direct and indirect impacts due to increased construction workforce traffic and construction truck traffic; negligible cumulative impacts because the number of workers needed for operations of all of these projects is modest compared to road capacities; Negligible direct and indirect impacts on traffic due to a low increase in operational traffic; Long-term adverse direct and indirect impacts on travel in the project vicinity because of BLM route closures; **`Visual Resources** Direct adverse impacts due to the very high magnitude of change to the visual landscape; indirect adverse impact of encouraging additional subsequent development of similar industrial character in the area; incremental contribution to adverse cumulative impacts; **Hydrology and Water Resources** Adverse, long-term direct and indirect impacts on surface hydrology due to a loss of on-site ephemeral drainages; adverse, long-term direct and indirect impacts on desert wash communities downstream of the project; potential adverse, long-term indirect impacts due to an increase in standing water onsite; incremental contribution to adverse cumulative soil erosion and storm water impacts within the Newberry Springs watershed; Negligible, long-term direct and indirect impacts on groundwater due to localized drawdown of the water table; negligible cumulative impacts due the size and capacities of the affected groundwater basins; the existing and proposed future uses of groundwater in the basins; and the relatively low water use requirements of the proposed action; Adverse, long-term direct and indirect impacts on California State jurisdictional waters; No direct, indirect, or cumulative adverse impacts on floodplains are expected due to the lack of FEMA designated floodplains onsite; detention/debris basins would completely retain flood flows resulting from a 100-year flood.

On the list of environmental direct and indirect impacts described above, the only main change between the Proposed Project and the Alternative Project (Preferred Project) would be on the impacts on "Cultural Resources," which are described as nonexistent or the Preferred Alternative in the FEIS: "No adverse effects to any cultural resources with the potential for listing on the National Register of Historic Places; no permanent long-term adverse direct, indirect, and cumulative impacts on other cultural resources due to construction activities and increased human access" (Bureau of Land Management 2010a, p. ES-17).

3.5 Public Concerns and Participation

The BLM's National Environmental Policy Act (NEPA) process provides opportunities for the public and agencies to participate and consult in the scoping of the environmental analysis and in the evaluation of the technical analyses and conclusions of that analysis. The scoping meetings occurred for the Calico Project in 2009 (Bureau of Land Management 2010a). During the scoping process, the BLM received letters concerning the project from members of the general public, representatives of community organizations, environmental organizations, tribes, and numerous federal, state, and local agencies (Bureau of Land Management 2010a). Some of the groups providing comment included Basin and Range Watch, The Wilderness Society, Wildlands Conservancy, National Resources Defense Council (NRDC), Society for the Conservation of Bighorn Sheep, United States Environmental Protection Agency, California Department of Fish and Game, and the Mojave Water Agency. Several areas of controversy regarding the Calico Solar Project were identified in this process, including, but not limited to: opposition to the placement of a large solar project on essentially undisturbed desert land; opposition to the overall number of renewable energy projects in the western United States; concern regarding the impacts of this project on biological, cultural, and visual resources; concern regarding the closure of BLM routes for renewable energy development; and concerns regarding the viability of the proposed solar technology.

3.6 Opinion of Environmentalists Regarding the FEIS and Project Location

The location of the project was probably the greatest issue for many of the conservation organizations. Excerpts (see Box 3.1) from descriptions of the project site in the FEIS indicate many of the potential issues the site presented (Bureau of Land Management 2010a, p. ES-9).

The conservation community argued that the proposed footprint fell on public lands containing high-quality habitat for threatened desert tortoises and for at least six other imperiled species. The footprint was also adjacent to the Cady Mountains

Box 3.1 Characteristics of the Calico Solar Project Location, According to the FEIS

The Calico Solar Project site is located in an **undeveloped** area of San Bernardino County, California, approximately 37 miles east of Barstow, California [...]

The project site contains a variety of vegetation types, including **four special or sensitive species**: white-margined beardtongue, crucifixion thorn, small-flowered androstephium, and Utah vine milkweed.

A total of **nine special-status wildlife species** were identified within or in proximity to the project site; these include desert tortoise, Mojave fringe-toed lizard, American badger, loggerhead shrike, Le Conte's thrasher, Bendire's thrasher, burrowing owl, golden eagle, and Swainson's hawk.

There are **335 cultural resource sites** within the Calico Solar Project area of potential effect, including 119 archaeological sites, 2 indeterminate rock feature sites, 206 archaeological isolates, and 10 historic built environment resources.

Four Wilderness Areas (WAs) and one Wilderness Study Area (WSA), Cady Mountains, are located in the project vicinity.

There are also **three areas of critical environmental concern** (ACECs) in the project area. The Pisgah ACEC, the Rodman Mountains Cultural Area, and the Ord-Rodman ACEC.

(which is habitat for Nelson's bighorn sheep and also contains nesting sites for golden eagles and prairie falcons), as well as sensitive public lands, such as a Wilderness Study Area, a Desert Wildlife Management Area, and areas of critical environmental concern designated by the BLM (Defenders 2011a; Defenders 2012b). In the view of organizations such as the Defenders of Wildlife, Sierra Club, and NRDC, Calico Solar Project's proposed location meant that harm it would cause to natural resources could not be mitigated: "With no way to fix this project given its current location, the harm it poses to imperiled wildlife, plants and fragile desert habitat seriously outweighs its benefits" (Defenders 2011b).

According to a report published by the Defenders of Wildlife in 2010, Tessera Solar and BLM (Defenders 2011a, p. 3)

> rejected consideration of a largely private land alternative identified by the California Energy Commission, including already-degraded agricultural lands and brownfields near the proposed project site. Instead, the developer chose to locate the project in an area the U.S. Fish and Wildlife Service has determined to be an important habitat linkage connecting several desert tortoise recovery units, which is essential for the long term recovery of the desert tortoise [...] even though FWS recommended prohibiting future renewable energy development on public lands in the Pisgah Valley in their entirety.

The Defenders of Wildlife view was supported by several organizations that strongly opposed the project, among them the NRDC, Center for Biological Diversity, The Wilderness Society, Audubon California, The Nature Conservancy, Sierra Club,

California Native Plant Society, Planning and Conservation League, Californian Coastal Protection Network, and Californian League of Conservation Voters (Defenders 2012b, p. 1).

3.7 Initial Project Approval

Tessera Solar's Calico Solar Project completed the joint State and Federal approval process in October 2010 and received California Energy Commission approval, as well as the BLM Record of Decision. Tessera Solar had a 20-year Power Purchase Agreement with Southern California Edison for the solar energy generated from the Calico Solar project (Renewable Energy Policy Group 2010). The approved project was to produce 663.5 MW and was located on 4613 acres in San Bernardino County. The amendment of the CDCA Plan allowed for the siting of the Calico Solar Project and related facilities, including 26,450 SunCatcher solar dishes, a 230- kilovolt Calico Substation, and a 2-mile, 230-kv transmission line. The BLM authorized issuance of a lease to Calico Solar for the Modified Agency Preferred Alternative (as described in the section above) (Bureau of Land Management 2010b).

Additional modifications to the Agency Preferred Alternative were made as the result of additional agency consultation and in response to public comments. The 6215-acre, 850 MW Agency Preferred Alternative for the Calico Solar Project was modified to 4604 acres and 663.5 MW. The modifications were not a result of the protests received during the Notice of Availability of the FEIS. Rather, they were a result of the California Governor's Office review and additional consultation with jurisdictional regulatory agencies (BLM 2010b). In the Record of Decision, the BLM explained the rationale for the decision (Bureau of Land Management 2010c, p. ES-2):

> These decisions fulfill legal requirements for managing public lands. Granting the ROW to Calico Solar, LLC, contributes to the public interest in developing renewable power from the Calico Solar Project to meet Federal and State renewable energy goals. The stipulations in the grant ensure that authorization of the Calico Solar Project will protect environmental resources and comply with environmental standards. These decisions reflect the careful balancing of the many competing public interests in managing public lands for public benefit. These decisions are based on comprehensive environmental analysis and full public involvement. The BLM engaged highly qualified technical experts to analyze the environmental effects of the Calico Solar Project. During the scoping process and following publication of the SA/DEIS, members of the public submitted comments that have enhanced the BLM's consideration of many environmental issues germane to the Federal authorization of the Calico Solar Project. The BLM, CEC, DOE, 001 and other consulted agencies used their expertise and existing technology to address the important issues of environmental resource protection. The BLM and 001 have determined that the measures contained in the FEIS, the Programmatic Agreement regarding the management of cultural resources, the Biological Opinion, avoid or minimize environmental harm to the maximum extent practicable.

3.8 Ownership and Project Changes

In December 2010, approximately 2 months after its approval, K Road Power acquired the Calico Solar Project from Tessera Solar for $3 billion. K Road, in contrast to Teressa Solar, would use Photo Voltaic, not dish engines, for 750 MW of the 850 MW (Prior 2010). K Road Managing Partner Gerrit Nicholas stated: "We are excited to move the Calico Solar Project into a financeable position, and we look forward to developing, constructing, and operating one of the world's largest solar projects. In addition, we are pleased to play a role in creating economic opportunity for the Barstow community and helping society wean itself off of carbon based power supplies through environmentally advanced technology" (SeeNews Renewables 2010).

In March 2011, Calico Solar developers filed a petition with the CEC requesting to modify the technology used in the Calico Solar Project from SunCatchers to PV. While the construction activities and methods proposed for the modified project would be similar to those described under the approved project, the intensity of the construction activities would be reduced, especially because PV does not require heavy equipment and a large workforce and vehicles (Basin and Range Watch 2013). K Road planned to leave much of the vegetation, as well as to review the Desert Tortoise Relocation/ Translocation Plan, the Final Burrowing Owl Relocation Management Plan, the Avian Protection Plan, Revegetation Plan, and Closure Plan. For the Mojave fringe-toad lizard, K Road would leave a small connectivity corridor open. The new project was being seen as far more environmentally friendly than Tessera's design, but still had some of the same flaws (Basin and Range Watch 2013).

An "Issues Statement" was released by CEC in April 2011 identifying unresolved problems with the project that the new applicant, K Road, needed to address. These issues were as follows: Tortoise Translocation, Rare Plant Surveys, State Protected Desert Washes, Final Drainage, Erosion, and Sedimentation Plan Needed, BNSF Rail Right-of-Way Access, and Glint and Glare Study. The CEC hoped to resolve all issues by September 2011 to allow construction to begin by the end of the year in order that K Road could still obtain a DOE loan guarantee and AARA grant that would cover 30% of construction costs (Basin and Range Watch 2013).

In July 2011, CEC held a status conference to hear arguments about whether the Energy Commission would also review and license the photovoltaic part of K Road"s Calico Solar Project or just the SunCatcher portion. The CEC previously only licensed thermal projects, but they agreed to be the lead agency for the PV part of the Calico Project. The Sierra Club argued that the Department of Fish and Game should be the lead agency for the PV portion of the project, and that environmental review should start from scratch. BNSF Railway generally agreed with the Sierra Club. K Road disagreed, saying DFG had only limited authority, and that it would only issue an Incidental Take Permit for threatened and endangered species (Basin and Range Watch 2013).

In October 2011, K Road Calico Solar applied to the BLM for a ROW grant amendment on public lands to include previously proposed changes in the project. Still in 2011, the BLM put out a Notice of Intent to prepare a Supplemental Draft Environmental Impact Statement for the Calico Solar Project. This document would analyze changes in the project that were not analyzed in detail in the first FEIS, including any additional site-specific impacts resulting from the change in technology, such as impacts to air quality, biological resources, cultural resources, water resources, geological resources and hazards, hazardous materials handling, noise, paleontological resources, public health, socioeconomics, soils, traffic and transportation, visual resources, waste management, worker safety, and fire protection, as well as facility design engineering, efficiency, reliability, transmission system engineering, and transmission line safety and nuisance (Basin and Range Watch 2013).

3.9 Challenges to Project's Approval and Amendments

Although CEC approved Calico in 2010 only after developers agreed to reduce its footprint to 4613 acres, and stated their belief that the revised configuration would reduce the impact on the desert tortoise by 79%, environmentalists were not satisfied. On October 2010, Gloria Smith, senior attorney for the Sierra Club, sent a letter to the Commission arguing that even a downsized project would prove devastating to protected wildlife (Woody 2010):

> It would result in the deaths of scores of threatened desert tortoises, result in the local extinction of the Mojave fringe-toed lizard, destroy an exceedingly rare desert plant, obstruct bighorn sheep movement, risk the survival of local golden eagles and impact burrowing owls, desert kit fox, and American badger [...] The mere fact that each of these species is even present on the Calico site is astounding.

In November 2010, Karen Douglas, chairman of the CEC, temporarily withdrew approval of Calico solar power due to protests regarding improper licensing of the project. In part, this was because an attorney for California Unions for Reliable Energy (CURE) had successfully argued that the CEC had not filed required written findings about Calico environmental consequences when it approved the project (Woody 2010). Douglas stated that the order for withdrawing approval did not mean that the Commission agreed with CURE's accusations (Cooler Planet 2010).

Soon after acquiring the project, K Road faced four lawsuits (CURE, Sierra Club, La Cuna de Aztlan, and BNSF lawsuits). Most of these lawsuits began while the project was under Tessera Solar ownership, as a result of its approval in 2010 (Basin and Range Watch 2013). Even though CURE had been a leader in the temporary withdrawal of Calico project by CEC, once the project changed ownership, the organization announced, through a joint statement with K Road in June 2011, that new collaboration between the agencies had been established (California Unions for Reliable Energy 2011a, b). Calico Solar and CURE issued a joint public pronouncement (California Solar Energy & California Unions for Reliable Energy 2011):

> Calico Solar, LLC, which is developing the Calico Solar Project, and California Unions for Reliable Energy (CURE) are pleased to announce that they have reached an agreement to resolve the outstanding issues raised by CURE before the California Energy Commission (CEC) concerning the Calico Solar Project.

As part of this collaboration with CURE, K Road would implement specific measures to avoid or minimize the project's environmental impacts as well as additional measures to protect worker health and safety during construction and operation of the facility. The K Road Calico Solar Project would use union labor and negotiate a Project Labor Agreement with local unions for the installation of both the planned PV and SunCatcher capacity at the project site (California Unions for Reliable Energy 2011a). Mike Mowrey, International Vice President of the International Brotherhood of Electrical Workers, stated the following (California Unions for Reliable Energy 2011a):

> CURE believes that the construction and operation of K Road Calico Solar will be of great benefit to the State of California […] The project was sited on largely disturbed land in a designated energy corridor and the project team has gone to great lengths to reduce and mitigate any environmental impacts. K Road Calico provides an example of how California can develop a low-carbon economy while simultaneously preserving valuable desert habitat.

Even after acquiring CURE's support, K Road still faced several challenges. In October 2010, immediately after CEC's approval, the Quechan Indian Tribe filed suit against the federal government over the approval of the Calico power plant. The tribe argued that the plant would harm the flat-tailed horned lizard, an animal proposed for endangered species protection and which is part of the tribe"s creation story (Woody 2010).

Still, regarding CEC's approval for the Calico Project in 2010, the Sierra Club filed a petition before the California Supreme Court challenging the decision. The Sierra Club's representatives were particularly worried about the area where the project was to be sited. The area was described by Sierra Club's senior attorney Gloria Smith as "a very unfortunate site" by virtue of being the habitat of endangered species such as the desert tortoises. In response, CEC spokeswomen Sandy Louey stated that CEC was "look[ing] forward to defending [their] position in court" (McBride 2011).

While the Sierra Club was filing the petition against the Calico Solar Project, the California Assembly was passing bill SB 266; this legislation gave CEC exclusive jurisdiction to approve changes on non-controversial solar projects without passing through the process of regulatory scrutiny. The five projects that were approved by CEC after SB 266's passage had been approved previously, but there had been changes made in the technologies since then from solar thermal to photovoltaic. The conservation coalition did not challenge any of these fast-tracked approvals; however, the Calico Project could not be included in this legislation, since the Sierra Club had filed a petition with the California Supreme Court challenging CEC"s prior approval of the project. On the one hand, Calico was not included on the list of projects that benefited by SB 266; on the other hand, the California Supreme Court refused to hear the petition or any other challenges to CEC decisions (Sahagun 2012).

In August 2011, the coalition of conservation groups made an appeal to Interior Secretary Ken Salazar, urging the Bureau of Land Management to move the Calico Solar Project to degraded lands. They also notified DOI of their intention to file a lawsuit against BLM and the U.S. Fish and Wildlife Service for approving the solar power facility, citing its impact on seven imperiled species (e.g., desert tortoise, burrowing owls, bighorn sheep). However, the coalition still hoped to work on an agreement with K-Road Power and BLM outside the courts if the developer agreed to move the project location (Defenders 2011b).

Alternative siting locations had been previously proposed by environmental groups, and the California Energy Commission found that the alternative sites would be good options for the project. However, the BLM and Tessera Solar rejected these locations. Like Tessera, K-Road Power had also been unwilling to consider a less environmentally damaging alternative location. The conservation community noted that degraded lands were available and could be used to produce the same amount of energy, but posed less risk to wildlife and the environment. The lands suggested by the coalition were mostly private lands, including brownfields, which are degraded agricultural lands and former industrial sites (Defenders 2011b). The following are some example statements regarding the proposal to change the project's location:

> Especially when there are ample disturbed and/or degraded sites throughout California that are appropriate for renewable energy development, selecting this specific land for the project has created a preventable conflict between the project's developers and Californians who want to preserve our state's native wildlife and landscapes. There is still time to get this right and we hope to work with K-Road Power to find an alternative location that has far fewer impacts on wildlife and desert ecosystems (Barb Boyle, Senior Representative for the Sierra Club; Defenders 2011b).
>
> The harm to imperiled wildlife, plants and fragile desert habitat caused by this project if it is built in this location seriously outweighs any benefits. And while we've worked successfully with a number of large-scale solar project developers to reduce their project's impacts on the environment and supported other projects with no changes, the impacts of this project are too great in its present location. As we have done successfully with other solar developers, we hope we can find a way to modify this project to reduce its impacts and permit it to proceed (Kim Delfino, Program Director for Defenders of Wildlife; Defenders 2011b).

In addition to the challenges from the conservation community and some tribal agents, the Calico Solar Project raised concerns from BNSF Railway. The railroad company owned a transcontinental line that ran through the proposed Calico site near Barstow. The line is a critical freight artery between the eastern U.S. and West Coast ports, according to BNSF. Traffic on the line reached more than 100 trains per day prior to the 2008 financial crisis, and currently is running at more than 80 trains per day during peak periods.

In building the project, K Road would no longer avoid washes or build detention basins, and BNSF worried about potential damages to their tracks (Basin and Range Watch 2013). The railway BNSF called the petition inadequate (California Energy Markets 2012, p. 2):

> Unlike the original SunCatcher proposal, the second petition to amend simply does not represent a specific plan with sufficient factual detail for meaningful evaluation," BNSF stated, adding that the recent petition does not address critical engineering issues related to storm water runoff and sediment transport.

In addition to the possibility of track damage or derailments, the railroad was also concerned about glint and glare from solar arrays that could affect the safety of trains running at speeds of 70 to 90 miles an hour.

In March 2012, Defenders of Wildlife, NRDC, and the Sierra Club filed federal litigation over the federal government's approval of the Calico Project (Sahagun 2012):

> BNSF Railway Co. has asked the CEC to dismiss a petition from K Road Calico Solar LLC, developer of the Calico Solar Project. K Road on June 26 filed a petition at the CEC to convert Calico, a 663.5 MW project initially proposed using solar-thermal technology, to solar-photovoltaic. It said it would use one of two PV technologies.

3.10 Approval of Bill AB 107

Despite concerns raised by local citizens, conservation groups, tribes, and the railroad, the California Assembly approved legislation (AB 107) that would allow a fast track for the evaluation and construction of the Calico Solar Power Plant (Newnet 2012). The bill AB 107 was introduced by Assemblyman Felipe Fuentes, who said its purpose was simply to clarify Calico Solar''s status under a bill approved on the prior year – SB 266 (see explanation for SB 266 process above). The rationale for AB 107 was that once the Supreme Court had refused to hear Sierra Club's case, there was no effective reason for omitting Calico from the projects that were approved under SB 266 (Sahagun 2012).

In 2010, the California Energy Commission had approved an earlier version of the Calico Solar Project, but after the project's approval, it had changed ownership, content, and scope. With the new legislation, the changes in the project—which were introduced when K Road Power became its owner—would not pass through additional local and state environmental studies, instead going straight back to the California Energy Commission (Defenders 2012). According to AB 107, San Bernardino County would lose its formal say in the approval process. The bill was aimed at "help[ing] the state meet one third of its electricity needs from renewable resources by 2020" (Newnet 2012).

Representatives from neighboring Kern County demonstrated concern that this bill would pose a threat to local control of such projects. Environmental groups were concerned that the commission was predisposed to overlook their concerns, since its mission is to help the state meet its 2020 goals for renewable resources (Sahagun 2012). Opponents of the bill planned on defeating it during Assembly committee; however, the process was fast-tracked and the bill was sent directly to the floor because it was considered urgent. The urgency to approve the bill relied solely on Calico Project developers' desire to meet a June 30 deadline to apply for approval from the Energy Commission (Sahagun 2012).

The conservation coalition (Sierra Club, Defenders of Wildlife, and NRDC) expressed frustration over the decision to send the bill to the floor without a hearing in committee. Kim Delfino, California Program Director for Defenders of Wildlife, commented (Sahagun 2012):

> It is disappointing to see California leaders put a corporation's interests ahead of the public's interest [...] Backroom deals between lawmakers and corporations that cut the public out of important decisions are the wrong way to build our clean energy future [...] It may not be the most beautiful place in the world, but if you are a tortoise or a Nelson's bighorn sheep, it is critical for the survival of your species.

Annie Notthoff, NRDC Director of California Advocacy (San Francisco and Sacramento), posted the following in her blog (Notthoff 2012):

> Passage of AB 107 will undermine efforts of California's environmental community to continue to do what California's elected leaders, including the Speaker of the Assembly and the Governor, have asked us to do — namely, to work hard with the solar industry to find ways to site large-scale solar projects in the desert while protecting unique and sensitive wildlife and other natural resources in order that the State can meet its aggressive and precedent-setting renewable energy goals.
>
> The legislature's action is a serious setback in the effort to plan for the best renewable energy future for our nation and California. The Calico solar project poses such a significant threat to the sensitive Pisgah Valley that it comes as no surprise that K-Road Power would push for legislation that short circuits any environmental review that would bring to light the project's true impacts. Unfortunately, California's lawmakers allowed a single developer to bypass environmental laws designed to protect the public's interest. This move is purely for the benefit of one company, does nothing to speed up our transition to clean energy, and will only lead to more controversy, costly delays and destruction of environmentally sensitive lands. The best way to ramp up renewable energy development is with an open, transparent process that engages local communities, Native American tribes and conservation organizations throughout the planning process.

Even a consultant and lobbyist to the energy industry, John White, had concerns about the process and potential public relations fallout: "This is a curious project to fast-track, given that it is controversial and facing litigation [...] it is going to make the renewable energy industry look bad in the end because it is an overreach" (Sahagun 2012).

3.11 New Proposed Amendments

In an attempt to ameliorate project concerns and opposition, K Road submitted a Petition for Project Technology Modification in June 2012 that proposed the following changes in the technology associated with the Calico Solar Project (California Energy Commission 2012):

- The proposed technology to be utilized for the modified project would be Photo Voltaic (PV) modules mounted on (1) horizontal single-axis trackers or (2) a

fixed tilt racking system. Fixed tilt PV panels may also be used on the site where terrain or other site constraints preclude the use of single-axis trackers.
- The Modified Project would still be constructed in two phases, Phase 1 would be located primarily south of the Burlington Northern Santa Fe (BNSF) railroad and Phase 2 would be located primarily north of the railroad.
- A new access route south of the BNSF railroad tracks to the most western portion of the project

In addition to changing the technology, K Road proposed siting modifications that would reconfigure the project layout to reduce the project footprint from 4613 acres to 3855 acres, as well as accomplish the following (California Energy Commission 2012):

- Remove the 5 to 1 tortoise mitigation lands located in the northern project site.
- Create a wildlife movement corridor through the center of the project.
- Reduce impacts to the White-Margined Beardtongue from four (4) distinct areas to one (1).
- Exclude 69 acres of Mojave Fringe Toed Lizard habitat. This total acreage includes 17 acres of the 21.4 total acres of high quality breeding habitat.
- Exclude the donated lands (Catellus Lands) within Section 17 from the site.

In 2012, Calico Solar still had no power purchase agreement with a utility, no financing, and no construction start date, and it faced a lawsuit by environmentalists (Sahagun 2012).

3.12 Project Closure

In June 2013, K Road canceled Calico Solar Project's review to amend the application from a solar thermal project to a photovoltaic project. Even with the inexpensive photovoltaic technology, K Road announced that the project was still not economically viable—especially after the close of federal subsidies for solar technology (i.e., Department of Energy and American Recovery and Reinvestment Act loan and construction grant). Even though tax credits still applied, the K Road decided to drop the application (Basin and Range Watch 2013). According to Calico Solar Project developers, the abandonment of the enterprise happened due to changing market conditions. Less than six months before the project was dropped, BrightSource Energy Co. had also announced that it would back off from its plan to implement a solar project in Riverside County because of cost issues (Zimmerman 2013).

An activist from Defenders of Wildlife wrote: "Though the company cited 'changed market conditions' as their reason for abandoning the Calico project, the truth of the matter speaks for itself—poorly sited projects are risky not only for the environment, but for the developers as well" (Sexton 2013). For Kim Delfino, Program Director for Defenders, this was "a victory in terms of stopping projects that are unfortunately sited, but it's too bad we have to fight renewable energy projects. If we put them in the right places we'll cheerlead for them. If they're in the wrong place, we'll fight them" (Zimmerman 2013).

Environmental organizations such as the Defenders of Wildlife have been working to identify appropriate locations through the Desert Renewable Energy Conservation Plan (DRECP)—a multi-agency effort to identify those parts of the desert that should be developed with renewable energy and those that should be set aside for wildlife and other natural resources (Zimmerman 2013). The region where Calico Solar Project would have been built was designated by DOI as a solar exclusion zone; this excludes the location as a site for other applications (Basin and Range Watch 2013). Due to its original CEC approval, Calico would not fall under the recent land-use policies restricting solar development from defined "exclusion zones." But its cancellation means that the lands proposed to be developed will now revert to the new rules (Montgomery 2013).

3.13 Lessons Learned

In the case of the Calico Solar Project, initial stakeholder involvement was triggered by concern about the desert environment and wildlife, underscoring that it is critical to maintain the pro-environmental image of solar energy in order to establish public support. The harm to wildlife that may be the result of poor siting policy has the potential to damage this image. Siting policy is already in place to protect wildlife; however, the scientific knowledge of the impacts of utility-scale solar on wildlife is generally lacking, and these policies thus may not be enough to prevent harm. Stakeholders, including environmental groups, must be given clear information that is easy to access. In addition to this, developers must take care to avoid perceived injustice. If the stakeholders believe they are not being given fair access to the siting process, or they feel that they are being forced to bear unfair costs associated with development, there is a risk of damaging future relationships and breaking down channels of communication.

References

Basin & Range Watch. (2013). *K Road withdraws calico solar project application. Basin & Range Watch.* http://www.basinandrangewatch.org/K-Road.html. Accessed 15 Aug 2016.

Bureau of Land Management. (2010a). *Final environmental impact statement and proposed amendment to the california desert conservation area plan for the calico solar (formerly SES Solar One) Project, San Bernardino County, California*. Barstow: BLM.

Bureau of Land Management. (2010b). *Record of decision for the calico solar project and amendment to the california desert conservation area land use management plan*. Barstow: BLM.

Bureau of Land Management. (2010c, October 26). Notice of availability. *Federal Register*, vol. 75. http://www.blm.gov/ca/st/en/info/fed_reg_archives/2010/10/calico_ROD.html. Accessed 15 Aug 2016.

California Energy Commission. (n.d.). *Calico solar project*. http://www.energy.ca.gov/sitingcases/calicosolar/. Accessed 15 Aug 2016.

California Energy Markets. (2012, July 20). Railroad objects to solar plant. http://www.stoel.com/files/CEM1190.pdf. Accessed 15 Aug 2016.

California Unions for Reliable Energy. (2011a, June 1). Joint tatement of Calico Solar Energy, LLC and California Unions for Reliable Energy. http://www.sbctc.org/cure/. Accessed 15 Aug 2016.

California Unions for Reliable Energy. (2011b, August 4). California unions for reliable energy partners with K road calico solar project. http://www.sbctc.org/cure/. Accessed 15 Aug 2016.

Cooler Planet. (2010, November 23). Controversial solar plant on hold. *SolarEnergy.net*. http://www.solarenergy.net/News/800251543-controversial-solar-plant-on-hold/. Accessed 15 Aug 2016.

Defenders of Wildlife. (2011a). The calico solar project: Solar done wrong. *Defenders of Wildlife*. http://www.defenders.org/renewable. Accessed 15 Aug 2016.

Defenders of Wildlife. (2011b, August 26). Controversial solar power plant challenged by conservation groups. *YubaNet.com*. http://www.yubanet.com/california/Controversial-solar-power-plant-challenged-by-conservation-groups.php#.Uo6SqeKAksY. Accessed 15 Aug 2016.

Defenders of Wildlife. (2012a). Re: AB 1073 (Fuentes) - Strongly opposed. *Defenders of Wildlife*. http://www.occnps.org/PDF/AB1073_OpposeLetter.pdf. Accessed 15 Aug 2016.

Defenders of Wildlife. (2012b). CA lawmakers pass bill short circuiting environmental review of controversial Calico solar project. *Defenders of Wildlife*. http://www.defenders.org/press-release/ca-lawmakers-pass-bill-short-circuiting-environmental-review-controversial-calico. Accessed 15 Aug 2016.

Kanellos, M. (2011, January 5). Dear environmental community: Please shut up. *Green Tech Media*. http://www.greentechmedia.com/articles/read/dear-environmental-community-please-shut-up. Assessed 15 August 2016.

McBride, S. (2011, January 5). With solar power, it's green vs. green. *Reuters*. http://www.reuters.com/article/us-solar-wars-idUSTRE7042ZR20110105. Accessed 15 Aug 2016.

Montgomery, J. (2013, July 1). K road gives up on calico solar project. *RenewableEnergyWorld.com*. http://www.renewableenergyworld.com/rea/news/article/2013/07/k-road-gives-up-on-calico-solar-project. Accessed 15 Aug 2016.

Newnet. (2012, May 11). California assembly approves construction of controversial calico solar plant. *News & Views*. www.newenergyworldnetwork.com/investor-news/renewable-energy-news/by-technology/solar/california-assembly-approves-construction-of-controversial-calico-solar-plant.html. Accessed 12 May 2012.

Notthoff, A. (2012, May 9). *Solar done wrong: AB 1073 is not the right path to a clean energy future*. Switchboard: Natural Resources Defense Counsel Staff Blog. http://switchboard.nrdc.org/blogs/anotthoff/solar_done_wrong_-_ab_1073_is.html. Accessed 15 Aug 2016.

Prior, B. (2010, December 29). Is the sun setting on tessera solar/stirling energy systems? *Green Tech Media*. http://www.greentechmedia.com/articles/read/another-nail-in-the-coffin-tessera-sells-calico-solar-project. Accessed 15 Aug 2016.

Reid, S. (2011, January). First-ever solar projets on public lands. *Sun & Wind Energy Magazine*. http://www.sunwindenergy.com/magazine

Renewable Energy Policy Group. (2010, January 22). Tessera solar's calico solar project achieves permitting milestone. http://www.energy.ca.gov/siting/meetings/2010-01-22_meeting/presentations/Tessera_Solar-Solar_Projects_2010-01-22.pdf. Accessed 15 Aug 2016.

Sahagun, L. (2012, May 11). Solar project in desert gets boost from california legislature. *Los Angeles Times*. http://articles.latimes.com/2012/may/11/local/la-me-solar-calico-20120511. Accessed 15 Aug 2016.

SeeNews Renewables. (2010, December 30). K road power acquires 850 MW calico solar project. *SeeNews*. http://renewables.seenews.com/news/k-road-power-acquires-850-mw-solar-project-in-california-123267. Accessed 15 Aug 2016.

Sexton, C. (2013, July 16). Lights out for calico solar. *Defenders of Wildlife Blog*. http://www.defendersblog.org/2013/07/lights-out-for-calico-solar/. Accessed 15 Aug 2016.

Woody, T. (2010, November 22). California puts Tessera solar plant on temp hold. *Reuters*. http://www.reuters.com/article/2010/11/22/solar-california-idAFN2229360120101122. Accessed 15 Aug 2016.

Zimmerman, J. (2013, June 24). BARSTOW: Calico solar plans withdrawn. *The Press Enterprise*. http://www.pe.com/local-news/local-news-headlines/20130624-barstow-calico-solar-lans--withdrawn.ece. Accessed 15 Aug 2016.

Chapter 4
Kittitas Valley Wind Power Project

Abstract The Kittitas County Commission opposed the wind power project due to local citizen opposition that feared the 350–410 feet tall turbines would obstruct the scenic beauty of the landscape, drive property values down, and lead to adverse wildlife impacts. However, regardless of local opposition to the proposed project, then Governor Gregoire supported the development of the wind park citing the 2006 citizen initiative establishing that 15% of all electricity used by major utilities comes from renewable sources by 2020. The Washington Facility Site Evaluation Council also supported the development of the wind farm and ultimately decided to overrule the county commission and local opposition groups. The case for development was based on environmental considerations to reduce greenhouse gas emissions, even if there were some local externalities.

Keywords Wind power project • Local opposition • State government over-ride

4.1 Introduction

> *The concern was renewable-energy developers would just throw up their hands and say, 'Washington is making it tough.' Thankfully, this long saga appears to be at an end.* (Marc Krasnowsky, spokesman for NW Energy Coalition (Cornwall 2007b)).
>
> *I fear this precedent will embolden energy companies to bypass local leaders and go to the governor to have projects imposed on communities.* (U.S. Rep. Doc Hastings, whose district includes Kittitas County, WA (Cornwall 2007b)).

Wind energy is an abundant renewable resource in many parts of the United States and can be used to generate mechanical power or electricity through the use of turbines. Modern wind turbines have four main components: a tower, rotor blades, the gearbox (where the generator is located), and the concrete foundation. On newer land-based models, the tower alone can be taller than the Statue of Liberty, with rotor blades adding half again as much to the overall height. Electricity is created when wind passes over the specially designed blades that can tilt and rotate for maximum efficiency. When wind reaches approximately eight miles per hour, the wind turbine in the gearbox starts producing power (see Fig. 4.1 for an example of this type of wind turbine). Wind turbines can be grouped together into a "wind

J.C. Pierce, B.S. Steel, *Prospects for Alternative Energy Development in the U.S. West*, Environmental Challenges and Solutions 4,
DOI 10.1007/978-3-319-53414-5_4

Fig. 4.1 Wind Energy Turbines (Brent S. Steel)

farm" or wind power plant that generates clean, non-polluting electrical power that can then be fed into a utility grid, like electricity from a conventional power plant (U.S. Department of the Interior 2016).

Although it is considered green, renewable energy, wind power is not completely free of controversy. As Cappiello (2013) points out, in a recent study, "federal biologists found that wind turbines had killed at least 67 bald and golden eagles since 2008." The government has since begun to enforce environmental laws aimed at wind energy facilities in order to protect birds, including "winning a $1 million settlement…from a power company that pleaded guilty to killing 14 eagles and 149 other birds at two Wyoming wind farms" (Cappiello 2013). According to the U.S. Department of the Interior (DOI):

> Although wind power plants have relatively little impact on the environment compared to fossil fuel power plants, there is some concern over the noise produced by the rotor blades, aesthetic (visual) impacts, and birds and bats having been killed (avian/bat mortality) by flying into the rotors. Most of these problems have been resolved or greatly reduced through technological development or by properly siting wind plants. (U.S. Department of the Interior 2016)

In addition to environmental concerns, such as impacts on wildlife, opposition to wind power, as with other alternative energy technology (AET) projects, has also been based on aesthetic, economic, and other concerns; these will be discussed in the case below.

4.2 Case Study Project

The Kittitas Valley Wind Power Project (KVWPP) is located in Kittitas County, Washington, approximately 12 miles west of Ellensburg. The central Washington wind farm currently consists of 48 Suzlon S88 2.1 MW wind turbines sited on approximately 5400 acres of privately and publicly owned forest and rangeland. The Suzlon S88 wind turbine is approximately 396 feet tall, including the 142-foot long rotor blades (coverage diameter of about 288 feet); it contains approximately 386 cubic yards of concrete in the foundation (Suzlon 2015). Sagebrush Power Partners and the Bonneville Power Administration (BPA) own substations, and Sagebrush has a permit to construct four more turbines. The energy generated is dispatched through the BPA transmission system (Energy Facility Site Evaluation Council 2013). This project not only generates local property tax revenue for Kittitas County, but also provides land-lease income to the owners of the rangeland on which the wind farm is sited (Pacific Power 2013).

The case for development was based on environmental considerations to reduce greenhouse gas emissions, even if there were some local externalities, i.e., "tradeoffs" (see Chaps. 1, 7 and 8). The Kittitas County Commission, however, opposed the wind power project because of concerns by citizens that the 350–410 feet tall turbines would obstruct the scenic beauty of the landscape, drive property values down, and lead to adverse wildlife impacts. On the other hand, Governor Christine Gregoire supported the development of the wind park, citing the 2006 citizen initiative establishing that 15% of all electricity used by major utilities must come from renewable sources by 2020. The Washington Facility Site Evaluation Council also supported the development of KVWPP and ultimately decided to overrule the county commission and local opposition groups. Although it was successfully developed, KVWPP was an extremely polarizing issue, given that local community leaders and citizens opposed the project, only to have it approved against their wishes.

4.3 Initial Application for Site Certification

In January of 2003, Zilkha Renewable Energy, through its wholly owned subsidiary Sagebrush Power Partners (which later became "Horizon Wind Energy"), applied to Washington's Energy Facilities Site Evaluation Council (EFSEC) for site certification of the Kittitas Valley Wind Power Project (Energy Facility Site Evaluation Council 2013). The company could have requested a permit through the Board of County Commissioners (BOCC); however, they instead chose to submit their project to EFSEC. According to Chris Taylor, project manager: "[the EFSEC's process] is a rigorous process to get all the issues out on the table and address them. Second of all, it is the appeal track issue. If there were an appeal, and we don't expect one, it would be expedited [...] We are wanting to be able to move forward on a timely fashion" (Ohrenschall 2003).

Zilkha's submission cover letter stated that the project would consist of 121 wind turbines and have a total capacity of 181.5 MW. In the initial project, wind turbines would be located on either side of Highway 97, between Ellensburg and Cle Elum. This site has high energetic wind resources, and is traversed by six sets of high voltage transmission lines. Zilkha wanted to interconnect the project with Puget Sound Energy's 230 kV Rocky Reach-White River transmission line, Bonneville Power Administration's (BPA) Grand Coulee-Olympia, or Columbia-Covington's 287 kV line, all of which pass directly over the KVWPP site. The project's output would be sold through a long-term power purchase contract (Zilkha Renewable Energy 2003).

While applying for KVWPP, Zilkha also submitted a proposal for the Wild Horse Wind Power Project, which would be located on 5000 acres sloping down from Whiskey Dick Mountain, about 10 miles northeast of Kittitas (Ohrenschall 2003). A leading opponent of KVWPP, Geoff Saunders, of Residents Opposed to Kittitas Turbines, called the remote Wild Horse site "far more appropriate" for wind development. He urged Zilkha to withdraw its Kittitas Valley proposal and instead seek county approval only for Wild Horse. Zilkha, however, proceeded with both ventures (Ohrenschall 2003).

4.4 SEPA Requirements and Environmental Impact Studies

When siting a new energy facility under SEPA (State Environmental Protection Act), EFSEC is required to hold a public information meeting in the county of the project's proposed location. To meet requirements, EFSEC hosted two scoping meetings on March 12, 2003: an agency scoping meeting and a public scoping meeting (Shapiro and Associates 2003). Approximately 25 people attended the agency meeting and another 150 people attended the public meeting (Energy Facility Site Evaluation Council 2007a). Attendees included representatives from Washington Department of Fish and Wildlife (WDFW); Washington State Department of Ecology (Ecology); Washington State Department of Transportation (WSDOT); Washington State Department of Community, Trade, and Economic Development (CTED); Washington State Attorney General's Office – Counsel for the Environment; Bonneville Power Administration (BPA); U.S. Fish and Wildlife Service; U.S. Department of Agriculture – National Forest Service; and Kittitas County Community Development Services (Shapiro and Associates 2003).

In these meetings, substantive comments not previously addressed by Sagebrush or identified by Shapiro and Associates' Initial Completeness Report (ICR) emerged. Key themes included noise; strobe lights; the health effects of light, dust, and noise; visual impacts; and seasonal variations, as well as effects on tourism; access to roads, recreation and commercial uses of DNR lands; variations on tax revenues and property values; employment of local contractors and craftsmen, as well as wages; impact on Kittitas's public services; and effects on certain animals and plants. A list of proposed setbacks suggested in the scoping meetings summary are as follows (Shapiro and Associates 2003, pp. 5–6):

- Turbines should be placed a safe distance from property lines and wildlife habitats, at least 1–2 miles.
- Suggested project setbacks would be 1500 feet from property lines, private roads, and wells, and at least 2000 feet from residences.
- The minimum setback distance should be 1500–2000 feet from property lines and at least one-half mile from residences.
- No single turbine should be sited closer than 1000 feet away from an adjacent, nonparticipating owner's property line, public or private road, or public pathway, or closer than 1500 feet from an existing residence.
- Setbacks should be at least 1500 feet from private roads and homes.
- Address project setbacks from homes, BPA lines, and Kittitas County road rights-of-way and provide a clear rational for proposed setbacks.

  Identify a larger setback for the proposed structures from the project perimeter; describe how close the turbines would be to the existing transmission lines.

Shapiro and Associates Inc. considered all these comments in the preparation of the Draft of Environmental Impact Study (DEIS) released in December 2003. A public hearing to receive comments about the DEIS was held on January 13, 2004 in Ellensburg, WA. EFSEC received oral comments, letters, and emails from tribes, agencies, organizations, and individuals. The DEIS analyzed offsite alternatives and an additional analysis was performed and documented in the Draft Supplemental EIS (DSEIS), released in August of 2004. Another public hearing was held for comments on the DSEIS in Ellensburg, WA (Energy Facility Site Evaluation Council 2007a). Comments delivered through letters and e-mails after the publication of DEIS and DSEIS were transcribed in volume two of the Final Environmental Impact Study (FEIS).

It is clear that most comments made after the DEIS revolved around the following issues: shadow flicker, noise, decrease in property value, and property taxes. The comments requested additional, more detailed analysis, updated data, or clarification (Energy Facility Site Evaluation Council 2007h). Key actors in this debate fell into five categories: individuals, local agents, state agencies, tribes, and other organizations. The position of each is described in Table 4.1, except for the individual opinions due to the large number of letters (Energy Facility Site Evaluation Council 2007h).

4.5 Final Version of the Environmental Impact Study

The final version of the environmental impact study (FEIS) was released in February of 2007. It was prepared based on information received from agencies, organizations, and individuals who submitted written and oral comments on the DEIS, as

Table 4.1 Perspective of local and state agencies, organizations, and tribal organizations

Local agencies:	**Other organizations:**
Kittitas County Community Development Services (representative: Derald Gaidos, Fire Marshal): Concerned about compliance with Kittitas County Public Works Criteria; suggested that agreements to provide for emergency services, fire, and EMS be established with the closest Fire/Hospital District or Department.	**Various Commenters, Kittitas C.A.R.E.S**: Supported the project and EFSEC's process; urged county to adhere to the deadline so that a decision was made in a timely manner.
Kittitas County Community Development Services (representative: Clay White, Planner II): Expressed concerns regarding the analysis of the projects impacts, stating that many of them that had not yet been properly studied.	**Thom Wallace, Northwest SEED:** Stated that they believe that the DSEIS was well done and seriously done, and that EFSEC was exceeding the minimum requirements regarding wild life preservation.
Kittitas County Department of Public Works (representative: Paul Benette, Director of Public Works): Expressed concerns with regard to transportation, access and safety on SR97, SR10, Bettas Road, and Hayward Hill Road; suggested further analysis on traffic impacts and air traffic communications.	**Sonja Ling, Renewable Northwest Project:** Believed project was well done; however, had some review comments on sections such as: no action alternative, vegetation, wetlands and wildlife, visual resources, communications, cultural resources and socioeconomics.
Yakima Regional Clean Air Authority (Representative: Les Omelas, control officer): Noted that local regulations apply to the project and a plan to control dust from site preparation, construction, and landscaping phases is required per YRCAA Regulation. Reminded that an asbestos survey and report should be filed with YRCAA for any demolition and renovation of any structure under that jurisdiction.	**Travis W. Misfield, Residents Opposed to Kittitas Turbines (ROKT):** Stated that "the DEIS as currently presented is legally insufficient for the following specific reasons, and we ask that you consider directing further study [...]" in several areas such as: EFSEC's legislative charge and responsibility under SEPA, proper definition of the proposal, discussion of alternatives (considered inadequate), completeness of information, etc.
Tribes:	
Yakama Nation (representative: Johnson Meninick, Program Manager): Noted that the project is within the lands of Yakama Nation that has high potential to contain cultural resources of interest to the tribe. The project would affect Lithosol environments, which are sensitive, difficult to replace, and contain a number of important medicinal and subsistence plants to the Yakama. Was also concerned with the riparian areas already degraded by other initiatives, but that they believe "have the potential to contain these plants again."	**Chris Taylor, Zilkha Renewable Energy**: Sent comments on behalf of Sagebrush based on analysis of their development team as well as consultants who were involved in the original studies for the Site Certification Agreement (SCA). Overall he believes the DEIS is thorough and addressed all issues raised during public and agency scoping process.

(continued)

Table 4.1 (continued)

State Agencies:	
Lisa Kelley, Washington State Parks and Recreation Commission: Stated that "while the State Parks certainly supports clean energy, there is concern this project may have substantial visual impacts to Iron Horse State Park's John Wayne Trail users. Wildlife viewing, especially recreational bird watching, will be impacted by the project from turbine noise disturbance, maintenance activities, and the reduction of habitat [,,,] Also, there could be significant mortality of birds, particularly passerines, accidentally flying into the turbines."	**Debbie Strand, CEcD, Economic Development Group of Kittitas County:** Sent additional information regarding taxes and stated "some new tax revenue is generated but not as great as originally projected. Additionally, taxes are lowered for all taxpayers in the county." She concludes "from an economic development perspective this is a good project for our community, and the Economic Development Group of Kittitas County supports the project due to the economic benefits it brings to Kittitas County."
Ted A. Clausing, Washington State Department of Fish and Wildlife: "We are generally satisfied with the information provided in the Draft EIS." Stated that they worked with Sagebrush to guarantee compliance with regulations since the initial application.	**Gerry Williams et al., Pautzke Bait Company, Inc.:** Wrote in favor of the project and urged EFSEC to grant the rezone as requested by Zilkha Renewable Energy. "We purchased our tract of land off Highway 97 […] Almost 30 years ago we knew we were buying marginal land and wanted to keep it that way. We have kept our land free of houses. Our neighbors, we suspect, were looking for cheap land to put homes on which they did and now their criticism is wind power will devalue their land?!"; "We believe that the wind generators have always been exciting to look at and are pleasing to the eye"; "Please consider the rights of all landowners who are in favor of this project as the small group of opponents continues to drown out common sense."
John Lane, Assistant Attorney General of the Counsel for the Environment: "The CFE takes no position in support or opposition of the KVWPP at this time." They sent comments that "seek to ensure the Final Environmental Impact Statement provides the public with the most detailed information possible."	**Keith Johnson, Kittitas Audubon Society**: Expressed concerns about deficiencies in the DEIS. Explained that the state provides little protection to avian species of concern to their organization. "Further, once a wind facility is installed, there appears to be no legal recourse to force changes to reduce mortality levels." Concluded saying that the DEIS notes possible formation of an Advisory Committee and requested to participate in it.
Stephanie Kramer, State of Washington Office of Archaeology and Historic Preservation: Offered comments regarding the cultural resources report and mitigating measures. Also stated that "because impacts upon historical [resources] are issues that are yet to be resolved, we will need more information to comment on the mitigating measures for visual impacts."	

well as from testimony and exhibits presented in the adjudicative hearings (Energy Facility Site Evaluation Council 2007e). There were major differences between the KVWPP proposed through the DEIS (which was target of several criticisms) and the one presented through the FEIS, which tried to address the concerns of most of residents and organizations.

Several unresolved issues identified in DEIS were further analyzed in FEIS. These include the following:

1. Wetland impacts: After the DEIS, Sagebrush applied for a Joint Aquatic Resource Permit Application (JARPA); it was determined that the activities were eligible for coverage under NWP 12, and Sagebrush agreed to comply to mitigate impacts.
2. Evaluation of economic impacts was incomplete in the DEIS, and there was no economic impact analysis for the 65-turbine proposal. In 2006, the Economic Development Group of Kittitas County prepared an updated economic analysis entitled *Economic Impacts of the Kittitas Valley Wind Project*, which was included in the FEIS.
3. In the DEIS, the indirect economic and environmental impacts associated with tourism generated by the project were unknown. The FEIS, however, addresses mitigation measures that could adequately reduce potential tourism-generated impacts.
4. In the DEIS, the indirect visual impacts on historical and tribal resources had yet to be determined, and clarification of the National Register of Historic Places (NRHP) eligibility status of the North Branch Canal tunnel had also not yet been received. In 2004, Lithic Analysts found that the project would not directly affect significant cultural resources in the project area. Tribal consultation with the Yakama Nation regarding the project's potential effect on tribal resources was ongoing during the publication of FEIS.
5. There was the potential for television interference in the Swauk Prairie area, and this issue was not properly acknowledged in the DEIS. In the FEIS, Sagebrush proposed field studies to measure whether the quality of television reception would worsen because of the project, and to develop a solution in cooperation with affected residents.
6. The DEIS did not include studies on radio interference; however, an analysis conducted by Sagebrush's consultants in 2004 demonstrated that KVWPP's turbines would not disturb radio operations in the project area beyond what is typical for suburban areas (Energy Facility Site Evaluation Council 2007a).

Although Sagebrush, in negotiation with ESFEC, agreed to several mitigation measures, some adverse impacts were considered unavoidable. These included adverse impacts to the Yakima Nation's tribal resources at or near the project site, as well as visual impact caused by altering the appearance of the rural landscape over a large area of the Kittitas Valley. Finally, the potential loss of bald eagle habitat was a significant, unavoidable adverse impact, even though the likelihood of fatalities among bald eagles was considered remote in the FEIS (Energy Facility Site Evaluation Council 2007a).

4.6 Second Application to the Board of County Commissioners

Even before the release of the FEIS by EFSEC, in August 2005 Sagebrush Power Partners presented a second application to Kittitas County. As part of its review of Zilkha's Kittitas Valley project, EFSEC, had directed the firm to pursue compliance with the county's wind farm regulations and "to make the necessary application for change in, or permission under, the Kittitas County land use plans or zoning ordinances, and make all reasonable efforts to resolve the noncompliance" process (Ohrenschall 2003). Projects such as the Desert Claim, proposed by EnXco, were already exclusively within the county's permitting.

The application was presented to the BOCC; it contained several revisions to the initial project and addressed concerns raised by the county and the public during the previous application and evaluation processes. The main proposed revisions to the initial project were as follows (Energy Facility Site Evaluation Council 2007a, p. 2):

- Reduce the number of turbines to a maximum of 80 units.
- Commit to not installing the smaller turbines, and instead using turbines with a net power output ranging between 1.5 and 3 megawatts (MW).
- Revise the turbine string layout to mitigate visual, shadow flicker, and noise impacts.
- Increase the setback from property lines of neighboring landowners without project agreements from 50feet to 541 feet beyond the tip of the blade at its closest point to the property line.

EFSEC staff reviewed the new application; a SEPA official determined that an addendum should be issued to update the project description and document the results of the analysis that evaluated the project's impacts after revisions. This addendum was issued in December of 2005 (Energy Facility Site Evaluation Council 2007a).

As Kittitas County reviewed the second application, Sagebrush attempted to mitigate the perceived impacts of the KVWPP. During the adjudicative hearings, Sagebrush agreed to independent environmental monitoring of the project's construction; to eliminate any demonstrated "shadow flicker" impacts in the area within one half mile of the project; and to provide mitigation measures, such that the project would produce minimal adverse environmental impacts (Energy Facility Site Evaluation Council 2007b). Because of residents' concerns with visual effects and shadow flicker (Energy Facility Site Evaluation Council 2008), Sagebrush further reduced the number of turbines to 65 units, and committed to shutting them down during periods when they might cause shadow flicker impacts to residences of nonparticipating landowners (Energy Facility Site Evaluation Council 2007a).

4.7 Application for Preemption and Disagreements That Emerged with BOCC

In February of 2004, Sagebrush applied for the first time for the preemption of local land-use and zoning ordinances of Kittitas County, stating that the county's requirements for land-use consistency duplicated EFSEC's process. In the view of Sagebrush, this was problematic, given that RCW 80.50.010 required no duplication in the siting process, and that decisions were made in a timely manner and without unnecessary delay. Their application, based on WAC-463-28-040, also stated (EFSEC 2004, p. 2):

> EFSEC regulations regarding local land use consistency contemplate reasonable efforts to resolve the issue in a traditional zoning context, not in a siting/permitting context. The county's lengthy and duplicative land use process is actually a project specific siting/permit process, which includes an invalid environmental impact statement review procedure. The process being demanded by Kittitas County, if followed, will undermine the EFSEC process as well as the direction provided by the legislature, and is in violation of EFSEC's preemption authority

In September of 2005, Sagebrush applied to BOCC, seeking the county's approval for all activities of KVWPP. Sagebrush had withdrawn their first preemption application while applying to the BOCC in 2005 (Kittitas County Board of County Commissioners 2006). However, in June of 2006, the county denied all permits to Sagebrush on the basis of the Growth Management Act (GMA) and the County Wind Farm ordinance KCC 17.61A.

The county's denial of Sagebrush's application, as stated in their Resolution 2006–90 (Resolution Denying the Kittitas Valley Wind Power Project Proposal Including Denial of Subarea Designation, Rezone Overlay, Development Agreement and Development Activities Permit), was based mainly on lack of compliance by Sagebrush with the county's requirement for turbine setbacks of 2000 feet from non-participating residences (Kittitas County Board of County Commissioners 2006). Sagebrush did not accept this specific requirement because, according to them, "if these setbacks are applied to this Project, it will result in further significant reduction in the number of turbines, rendering the project unviable for the Applicant's commercial purposes" (Energy Facility Site Evaluation Council 2006, p. 3).

The disagreement regarding the distance of turbines from residences could not be negotiated. On one hand, BOCC argued that Sagebrush was not willing to collaborate with local policies and lacked faith in the local process. In their motion against EFSEC's recommendation to accept KVWPP, the BOCC stated that the applicant acted in bad faith and refused to negotiate (Energy Facility Site Evaluation Council 2007g). On the other hand, Sagebrush repeatedly stated that they tried their best to comply with the county's requirements (Energy Facility Site Evaluation Council 2006, p. 3):

> The Applicant disputes those portions of the County's Findings of Fact related to County Resolution 2006-90 which suggests the Applicant was unwilling to discuss project revisions. The Applicant, during the public hearings and in written communication with the

> county cited at the end of this section, also proposed technical means for resolving the shadow flicker issues after construction and operation if it is established that an actual significant impact exists and expressed the company's willingness to address road restoration and other project-related topics raised in the record. It was only the county's requirement that 2000 and 2500 foot setbacks be imposed – after the Applicant had dramatically reduced the number of turbines, proposed quarter-mile setbacks, and identified technical solutions to the shadow flicker issue – that did not lend itself to further negotiation.

Sagebrush initially proposed a setback of 1000 feet, and later proposed 1320 feet, neither of which was accepted by the county (Kittitas County Board of County Commissioners 2006; Energy Facility Site Evaluation Council 2006). BOCC did not accept Sagebrush's assertion of economic unviability because the company failed to explain it or present proof. The BOCC also stated that, in the company's proposal, over 40 nonparticipating landowners would experience shadow flicker, and over 20 of them would be located within a half mile of proposed turbines (Kittitas County Board of County Commissioners 2006).

Besides this, opponents argued, the project area was designated under the comprehensive plan as rural lands,[1] which are not adequate for wind farms, as stated in KCC 17.61A (Kittitas County Board of County Commissioners 2006). Sagebrush, in its defense, argued that "in accordance with KCC 17.61A, the entire county is inconsistent for wind farm use because wind farms are not designated as a permitted use in any of the county's zoning classifications" (Energy Facility Site Evaluation Council 2006, p. 9).

Due to BOCC's denial of their application, Sagebrush submitted a second request for Preemption in June of 2006. In this request, Sagebrush stated that they had been "committed to making a good faith effort to respond to county inquiries and resolve perceived noncompliance issues" (Energy Facility Site Evaluation Council 2006, p. 1). They argued that this could be demonstrated by their willingness to "completely redesign the site from the originally proposed layout, reducing the number of wind turbine generators from approximately 121–65 and resulting in an electrical power generation capacity loss of approximately 60 MW" (Energy Facility Site Evaluation Council 2006: 2). In addition, Sagebrush pointed out that they had offered a "redesign of the project which guaranteed quarter-mile setback buffers from the small number of adjacent landowners not having leases with the Applicant" (Energy Facility Site Evaluation Council 2006, p. 2).

[1] Kittitas County was especially concerned about rural lands. In 2007, they were still working on having a designated land area in the east side of the county that would be compatible for wind farm development. The new zone, according to County Planner Joanna Valencia, "would not replace the existing county wind farm resource overlay zone ordinance that now allows wind power developers to attempt to gain project approvals for nearly any rural area of the county. Wind farms proposed in that zone could take advantage of an alternative approval process. A variety of environmental studies will be required, along with approval of a site plan and development agreement from county commissioners. Hearings, however, would only be before the commissioners and not the state" (Johnston 2007).

4.8 Recommendation to the Governor and Motions for Reconsideration

The Energy Facility Site Evaluation Council is the state agency that recommends to the governor whether a new major energy facility should be sited in the state of Washington (Energy Facility Site Evaluation Council 2013). On March 27, 2007, through Order No. 826, the EFSEC recommended the approval of a Site Certification Agreement to construct and operate up to 65 wind turbines that would generate between 100 and 180 megawatts (MW) of wind power (Energy Facility Site Evaluation Council 2007b). This recommendation stated that the project was consistent with Kittitas County's comprehensive plans. It added that the project respected zoning code and regulations, except for the 35-foot height restriction in the forest and range zone and the Wind Farm Overlay Ordinance (Kittitas County Code Chapter 17.61A), which required BOCC to approve a wind farm project. "Therefore," EFSEC wrote "determining that the county's siting ordinance duplicates EFSEC's site evaluation process and usurps this Council's statutory authority, the Council recommends preemption of Kittitas County's Wind Farm Overlay Ordinance as well as the height restriction" (Energy Facility Site Evaluation Council 2007b, p. 2).

The final project that was recommended to Governor Christine Gregoire involved the construction of 65 wind turbines that would permanently occupy 108 acres of land on either side of US Route 97. It also included approximately 13 miles of new roads and improvements to roughly 8 miles of existing roads, 23 miles of underground and 2 miles of overhead electrical power lines, two new substations, a 5000-square-foot operations and maintenance facility, and up to five permanent meteorological towers (Energy Facility Site Evaluation Council 2007g). In its press release, EFSEC also stated that during its construction, the project would generate $5.8 million in direct income to Kittitas County; once completed, it would generate over $1.5 million in property tax annually (Energy Facility Site Evaluation Council 2008).

In early 2007, KVWPP was generating some controversy, as reflected in local newspapers. Editorials on the topic from different points of view began to appear (see the Box 4.1 for excerpts from example editorials about KVWPP).

Box 4.1 Kiittitas Valley Wind Project Editorials

"An Inconvenient View," *The Seattle Times* (2007a):

"Most objections have to do with how the turbines will alter the landscape. The sparsely populated area is developing with vacation homes nearby, and some local officials are concerned about tourism."

"The landscape will be different, to be sure. But the turbines can contribute to staving off starker changes."

(continued)

Box 4.1 (continued)

"Governor Should OK Revised Wind Farm for Kittitas County," ***Yakima Herald-Republic*** **(2007):**

"About the only thing as certain as wind in Kittitas County was the inevitable opposition from some area residents who don't like the thought of a wind farm going into their 'backyards,' using the loosest definition of that term."

"We understand their concerns, but they appear to have been sufficiently mitigated by developers. The project should move ahead."

"Horizon has already made significant changes to soften impact of the project, including cutting the number of wind turbines nearly in half—from 121 to 65. That's no small concession."

"[O]ne of the best selling points for the profitability of this wind farm is that it is in close proximity to a major corridor of power transmission lines that already cuts through the area. That can't be too pleasing to the eyes either."

"Unfortunately, we don't live in an ideal world and concessions and compromises are going to have to be made as the state embarks on serious energy initiatives."

"State Energy Battles are Blowing in the Wind," ***The News Tribune*** **(2007):**

"In the case of proposed energy facilities, however, local opposition alone **cannot** be allowed to kill projects that provide a greater good for the state."

After EFSEC's recommendation, the BOCC sent to the governor and to EFSEC a motion to reconsider the project as approved. They explained that BOCC has a process to review wind power applications that involves a SEPA process, an environmental impact statement, public hearings, and relevant public testimony. They stated that the Wild Horse Wind Power Project was an example of how their process worked and how effective and efficient it was. They argued that applying to EFSEC was optional on their processes, and that was normally done for the applicants' internal business purposes. Furthermore, they indicated, "in reviewing the record from EFSEC, we feel that the EFSEC Board members have erred in their review of the relevant issues. There were two issues which caused the BOCC to deny the KVWPP application" (Kittitas County Board of County Commissioners 2007, p. 1).

In the motion presented before the EFSEC, BOCC stated that the EFSEC's recommendation was without merit because of the following reasons:

1. Lawfully adopted GMA planning regulations to consider siting of wind farms and EFSEC had made local planning irrelevant by considering themselves as the GMA decision maker for Kittitas county, as stated in Order 826 (Energy Facility Site Evaluation Council 2007b, p.16).

2. EFSEC ignored the FEIS: independent consultants prepared the results of the FEIS, but EFSEC based its decisions solely on the applicant's paid expert testimony, which contradicted the FEIS, as stated in Order 826 (Energy Facility Site Evaluation Council 2007b, p. 31).
3. The setbacks sought by Kittitas County were based upon objective analysis: the BOCC had set a half-mile setback from adjacent homes based on the FEIS in order to change the level of impact on residents from significant to moderate.
4. The applicant acted in bad faith and refused to work with the county: the applicant counter-proposed a setback of 1320 feet and refused to negotiate regarding the Kittitas County land use process. According to the motion, the applicant gave an ultimatum to the County by saying: "at this point you could vote thumbs-up or thumbs-down on this project. The application is before you" (Kittitas County Board of County Commissioners 2007, p. 5).
5. The EFSEC's recommendation should be either of denial based upon rejection by the applicant or denial based on bad faith of the applicant: The applicant rejected BOCC's proposal of a half-mile setback saying that this would make their project unviable; they said 1320 feet was the greatest setback they could accommodate. However, later on, they agreed to EFSEC's setbacks greater than 1320 feet, which indicates bad faith. Therefore, Sagebrush "has not shown a good faith effort in the local process and their application should be remanded back for further consideration at the county level or denied by the Governor, upholding the local land use decision making body as required by GMA" (Kittitas County Board of County Commissioners 2007, p. 2).
6. EFSEC's recommendation conflicted with its own regulations: EFSEC's criteria required that the applicant demonstrate that there are no other alternative locations in the county. Although there are documented alternative sites in the county, EFSEC accepted compliance by stating that the applicant did not have control of property at any alternative site. BOCC's motion stated: "if this were the case, a proponent of a cogeneration power facility could site such a facility in GMA designated agricultural lands of long term significance, even if the county has heavy industrial zones lands suitable for such a project, by simply choosing to lease the agricultural land rather than the industrial land" (Kittitas County Board of County Commissioners 2007, p. 6).
7. EFSEC's recommendation conflicted with Initiative 937 and the interests of the state: EFSEC recommendation of preemption would override local land use and use up the limited capacity of the County's power lines. Besides, there was no proof that the power generated by the new facility would be sold to Washington residents; there were rumors that the power would be sold to California residents instead. Initiative 937 called for more renewable energy in Washington and exporting this renewable energy contravenes the Initiative's goal and the Governor's statements on renewable energy and energy independence (Kittitas County Board of County Commissioners 2007).

The motion concluded with the following statement (Kittitas County Board of County Commissioners 2007, p. 5):

> The decision of EFSEC ignores GMA planning goals. The applicant has failed to make a good faith effort to resolve for non-compliance. The applicant has failed to demonstrate that alternate locations within the County have been reviewed and found unacceptable. The applicant has failed to demonstrate that this project meets the interest of the State. The applicant request for preemption must be denied.

Directly addressing Governor Gregoire, BOCC representatives Alan Crankovich, David Bowen, and Mark McClain stated their mistrust of her judgment, based on her previous indication that deference must be given to the local legislative authority and on her understanding of the full implications of preempting the local land use authority on all land-use decisions. They concluded (Kittitas County Board of County Commissioners 2007, p. 2):

> We trust that when you review the record you will see the project does not meet the mitigation requirements of the EIS, that you understand preemption of a local land use decision would be inappropriate, and the applicant did not provide sufficient information to the process from which the BOCC could approve the project in good faith.

Kittitas County Commission Chairman Alan Crankovich, in an interview with the *Daily Record*, said that the county had done all it could do, at this point, to exercise local government control over county land-use decisions involving wind farms. He also mentioned that the state's Growth Management Act and other laws give local government decisions priority over state government, and said: "if she ultimately approves the wind farm, well, we'll have to explore other options [...] It's a huge decision" (Johnston 2007).

Another letter sent to EFSEC requesting reconsideration of Order 826 was written by resident Ed Garret, on behalf of himself and his wife, Rosemary Monaghan. Mr. Garret and Ms. Monaghan were members of a group called Residents Opposed to Kittitas Turbines (ROKT), but wrote their letter independently. Regarding compliance with the interests of the state, they argued: "there was no mention of other 5–7 wind projects being built in eastern Washington or in the planning state [...] that is obvious negligence. This alone proves preemption is not warranted for 'the good of the State.' There is no guarantee that the small amount of power will even benefit the citizens of Washington State. So much for the I-937 'mandate' which was NOT overwhelming" (Garret and Monaghan 2007, p. 1).

Regarding the residents' final situation, Mr. Garret stated, "The document neglects to say that there is 1 RESIDENT in the area that has a signed agreement with Horizon, while there are 16 others who are non-participating. Property line issues are glossed over for numerous (hundreds) properties who can build, and approved land use, but are holding off until this is decided." In addition, Mr. Garret argued that the public was being marginalized in the process: "your findings TOTALLY disregard our comments. Not one footnote references anything said by the public. This whole process is made political and for 'the good of the state'" (Garret and Monaghan 2007, p. 2). Garret also mentioned conflict of interests in the process of evaluation and recommendation of the KVWPP. He stated that the chairman of EFSEC was working with western Democrats to

> … write out participation of the public because, as he said in the Adjudication Hearing, the public is drawing this out and slowing up 'his process'. […] to have a standing Chairman of a recommending Council, while a crucial decision is being made, trying to undermine the concerns of private citizens to force an unwanted and unwarranted project on a whole county is egregious. Being appointed by the Governor (Locke), he should be forced to resign for conflict of interest and expert contact violating the Doctrine of Fairness (Garret and Monaghan 2007: 2).

Conflicts of interests were also said to be present in the representation of the council by the Department of Natural Resources and the Washington Department of Community, Trade and Economic Development. According to Garret, "these agencies are, or will be receiving funds from this Applicant wind developers and we still cannot conceive that this is not a conflict of interest" (Garret and Monaghan 2007, p. 2).

In response to the comments received regarding Order 826, EFSEC issued the Order 827 (on motions for reconsideration of Order 826 by Kittitas County as well as Mr. Ed Garret and Ms. Rosemary Monaghan). This document contained replies from Sagebrush to the two reconsideration requests, as well as a reply from Kittitas County for the motion received from Mr. Garret and Ms. Monaghan. In response to the requirement for GMA compliance, EFSEC "determined and reaffirms that the 'Overlay Ordinance' conflicts with and duplicates the Council's siting authorities contained in RCW 80.50" (Energy Facility Site Evaluation Council 2007c: 2).

Regarding the issues raised by Kittitas County on the visual impacts and the accuracy of the FEIS, EFSEC acknowledged that not all visual impacts could be eliminated, but also re-affirmed that its setbacks would provide sufficient mitigation. Regarding the accusation of bad faith on the part of the applicant, EFSEC quoted Order 826 saying that Sagebrush "expended significant effort to navigate the County's permitting process and that these efforts to resolve the land use noncompliance issues were made in good faith" (Energy Facility Site Evaluation Council 2007b, p. 21; Energy Facility Site Evaluation Council 2007c, p. 2). Regarding the existence of alternative sites, EFSEC stated that the Council used five reasonable criteria in alternative-site evaluation and restated that the proposed site was the best available undeveloped wind resource remaining in Kittitas County. Regarding the contraposition to I-937, Order 827 indicated: "the Council agrees with Sagebrush Power Partners and the Department of Community, Trade, and Economic Development's responses to the Motion for Reconsideration. The project does not conflict with Initiative 937" (Energy Facility Site Evaluation Council 2007c, p. 3). EFSEC concluded as follows:

> [P]ursuant to WAC 463-30-335 and RCW 34.05.470, the Council has considered the Kittitas County, and Mr. Garrett and Ms. Monaghan's motions for reconsideration. The Council finds and concludes that neither motion raises any factual or legal arguments that the Council has not already heard during the adjudication and in post-hearing briefs, deliberated upon, and discussed in Order No. 826 and the appended Draft Site Certification Agreement. No basis has been provided to justify any changes in the Council's findings, conclusions, or recommendation. Hence, the Council denies the Kittitas County's and Mr. Garrett and Ms. Monaghan's motions for reconsideration" (Energy Facility Site Evaluation Council 2007c, pp. 3–4).

In addition to Mr. Garret, other citizens also appealed to the Governor by posting an online petition that gathered 371 signatures. The goal of the petition was to gather 1000 signatures by September 30, 2007. The "Petition to Deny KVWPP and Desert Claim Wind Power Project" (see Box 4.2) was posted on the Petition Site by Cliff and Tina Sands (Sands and Sands 2007).

Some residents posted letters addressing the governor on the Internet. With the authorization of the writers, Linda Schantz, a resident opposed to KVWPP, posted some of these letters on her farm's website. Table 4.2 shows excerpts from some of the letters (Schantz 2007).

An article in *The Seattle Times* (see Box 4.3) attempted to show the positions of residents both in support of and opposed to the project, environmental groups, and County officials (Cornwall 2007a, b):

Box 4.2 Petition to Deny KVWPP and Desert Claim Wind Power Project

"The Governor of the State of Washington has before her a decision whether to override local county government's careful consideration and denial of wind power projects that will greatly impact many neighboring residents. These projects are being proposed in the northwest end of the Kittitas Valley – the Kittitas Valley Wind Power project (KVWPP), and the Desert Claim project applied for by EnXco.

Horizon Wind Energy (a Portuguese owned company) is trying to have their wind power project approved by the non-elected Energy Facility Site Evaluation Council (EFSEC) by pre-empting a Kittitas County land use decision. In the local permitting process, Horizon walked out when minimal setbacks were requested to reduce the impact on the residents of the community.

This same company planned and built a very large wind power facility (the Wild Horse Wind Power Project owned by PSE) using the very same local land use permitting process. This is an issue of proper siting of wind power projects so they do not negatively impact the community on whom they are being imposed.

We are not suggesting that EVERY proposed wind power project should be built or that NO wind power projects should be built. The issue is siting and the protection of residents that will be forced to live next to these types of projects.

We are asking for signatures to request that the Governor deny the pre-emption of the County's carefully considered denial of the applications to build the KVWPP and Desert Claim wind power projects, these two specific sites."

Table 4.2 Citizen opposition to KVWPP posted by Linda Schantz

Posts by Linda Schantz	***Posts by Residents of Ellensburg***	***Anonymous Posts***
"Last July, you helped dedicate the Wild Horse Wind Farm in east Kittitas County. Many did not oppose this project because it was sited in an appropriate location where very few homeowners lived nearby. The handful of people who lived within a mile or two were in support of the project. I wanted to remind you of your statements that day and ask you to consider them in making a decision on the KVWPP. Community Support was a key factor in approving this [Wild Horse Wind Farm] project. Hundreds in this community from the beginning have not supported this project. There have been hundreds of letters opposing the project and hours of public testimony where the majority of the opinions were opposing the location of the Wind Farm."	"EFSEC has totally ignored the preponderance of public commentary on this project. They summarize it on page 25 of their Council Order No 826 stating that: '… some rural residents emphasized the undesirable nature of having one or more wind turbines as neighbors, others stated their preference for a wind farm as opposed to seeing agricultural or rangeland further subdivided…' This is a gross misrepresentation of the ratio of people opposed to the project versus people who were in favor. Your committee is misleading you on this point, diminishing the scope of the community opposition. In meetings with audiences of typically 100–120 residents, the majority of the public verbal and written commentary was against the siting of wind power."	"Wind is primarily an energy resource that makes relatively little contribution to meeting system peak loads. Even with large amounts of wind, the Northwest will still need to build other generating resources to meet growing peak load requirements. […] When wind energy is added to a utility system, its natural variability and uncertainty is combined with the natural variability and uncertainty of loads."

(continued)

Table 4.2 (continued)

<table>
<tr><td>"If we are going to have a model here, then let's be a model. […] In order to be a model, wind farms need to be sited properly."</td><td>"You were quoted in our local paper as having given credence to the proximity to residents when considering the Wild Horse wind project. We beg of you to listen to the people, as your committee has not done, and deny the KVWPP application."</td><td>"Don't position BPA as some sort of benevolent renewable energy 'sugar daddy.' BPA will be more than happy to integrate your beloved wind power, but it will not be with hydroelectric resources and you will soon be paying dearly for their services. Any theoretical tax benefit county residents may receive will be more than used up paying higher utility bills."</td></tr>
<tr><td>"This project is being shoved down Kittitas County and its residents' throat. The county said no, so Horizon ran to EFSEC to have them say yes. They also had the time to create a market for their product by lobbying for renewable energy initiatives that don't count hydro-electric power as renewable! Governor, this is not bottom up, this is shoving it down our throat!"</td><td>"We and many others have invested all of our equity, blood, sweat and heart into our home here. We have done this at great sacrifice with the objective of living where it is quiet and where we may enjoy a rural lifestyle. […]. Such projects must be restricted to remote areas where their impacts on the population and land values may be reduced."</td><td>"This is not the only location in the county where a wind farm can be built. EnXco has a proposed Kittitas County wind farm project already before EFSEC, and Invenergy has begun discussion with the county about another wind farm near the Wild Horse project. The KVWPP developer may not have other options here, but that is only because its competitors have locked up other county locations." (Anonymous)</td></tr>
<tr><td colspan="3">Anonymous, quoting the "Northwest Wind Integration Action Plan" written by the Northwest Power and Conservation Council</td></tr>
<tr><td>"This increases the need for flexible resources such as hydro, gas-fired power plants, or dispatch able loads to maintain utility system balance and reliability across several different timescales […]"</td><td>"The reduction in our home value would make it impossible for us to sell and relocate, so we would be forced to remain here and endure the constant intrusion of the industrial spinning towers. There has not been any indication that the corporations who are building these projects will be compensating us for our losses. Nor are they making the necessary effort to reduce the impact with appropriate setbacks as were requested by the County Commissioners."</td><td>"The local decision to deny this project was made by elected officials who represent county residents. The county's authority must not be pre-empted by the state except in the most exceptional circumstances. There are no exceptional circumstances here."</td></tr>
</table>

(continued)

Table 4.2 (continued)

"Current tariff and rate provisions may not be sufficient to recover the costs of wind integration in all cases. Cost recovery and allocation is a threshold issue for each control area. [...] The cost threshold has been reached at BPA [company that would collaborate with Sagebrush to distribute the energy]. By the end of 2007, BPA may have as much as 1500 MW of wind in its control area. In 2007, BPA expects to revise its approach to recovering and allocating wind integration costs through formal rate case proceedings."		"When you approved the Wild Horse project you made commitments that you would not force projects on unwilling communities or pre-empt elected officials. You were reported in a local newspaper, the Daily Record, as saying, "Gregoire asked how close people lived to the project area and said she wanted to make sure Wild Horse would have no significant impact on people. She said her inclination is to have wind farms gain community support from the bottom up, not the top down. It was important in her approval that pre-emption didn't occur with the Wild Horse project, she said." We ask that you honor those commitments. Denying this project will not prevent more wind farms from being built in the county. But they will be built in more suitable locations, and with the support of residents and local government."

Schantz (2007)

4.9 Governor's Request for Reconsideration of Article I(C) (7)

On June 22, 2007, Governor Gregoire requested that EFSEC reconsider the recommended setback for some of the turbines. In a letter to EFSEC, she stated: "My directive to EFSEC that the Agreement be reconsidered is solely focused on the need to determine on this particular Project whether additional setbacks beyond the four times height (4xh) requirement for non-participating landowners are achievable while allowing the Project to remain economically viable. EFSEC's recommendation, inclusive of attachments, is silent on the issue." The specific Article of the Site Agreement that the Governor was referring to [Article I (C)(7)] are summarized below (Energy Facilities Site Evaluation Council 2007i: 11).

Turbine Setbacks, Article I(C)(7)

Turbines shall be set back from existing built elements as follows:

- Distance from residences, existing as of January 13, 2003, of adjacent landowners without signed agreements with the Certificate Holder: four times the maximum turbine tip height (i.e. 330-foot turbine = 1,320-foot setback; 440 = 1,640-foot setback).

Box 4.3 "An Ugly Battle Over Clean Power," *The* Seattle Times (2007a, b)
Testimony of resident opposed to the project:

"Sandy Sandall is a retired Boeing mechanic who moved into a log home on a forested ridge in 1997 because of the quiet and the view. In his interview, he says he does not oppose wind power, in fact, he even gets his electricity from solar panels and from his own relatively tiny wind turbine mounted on a 50-foot pole. But he considers that rows of white massive white poles, topped with blinking lights and propeller blades that make a "whoosh" audible from several hundred yards away — and that is another thing. "I'm 72 years old. I came out here to be alone [...] I have no objection to alternative energy. This is just not the place."

Testimony of resident who supports the project.

"Mike Genson, a retired local schoolteacher, agreed to let Horizon build seven turbines on his land. He said he was skeptical when the company first approached him. But after he visited another wind-power project near Walla Walla, he decided that he didn't mind the way it looked or sounded. Besides, he said he likes the idea of providing clean power and that the income — possibly $30,000 a year or more — could allow him to keep the property, where he lives part time. "If we don't get the wind farm, we're going to be forced into gradually letting this go to development."

Environmentalists and Business Groups.

"The article stated that the fight regarding KVWPP produced some unusual alliances, namely, between environmentalists and business groups, including Kittitas Valley's economic-development organization, who are backing Horizon's bid. "The fact that someone doesn't like the looks of a wind turbine, I have some sympathy for them [...] But that's just not as important as dealing with the climate and our air." (Sara Patton, executive director of the NW Energy Coalition, a Seattle group).

Kittitas county officials.

"Kittitas County officials urged her to reject the EFSEC ruling. Darryl Piercy, head of the county's Community Development Services department, said it was a test of whether local development decisions would be honored by the state. "This is troubling to think that that local process can easily be ignored," he said.

- Distance from residences of landowners with signed agreements with the Certificate Holder: maximum turbine height (also called "blade tip height").

For purposes of this article, "existing built element" is defined as those built elements constructed as of January 13, 2003, or properties with vested rights to build as of January 13, 2003. For purposes of this article, "residence" means the primary physical structure on a residential lot utilized as a single family home; as used herein, "residence" includes the entire structure within the main walls and the eaves of the roof, but does not include uncovered decks, uncovered patios, or outbuildings.

Distance shall be measured horizontally from the centerline of the turbine tower to the outermost envelope of the built element considered. Turbines shall be set back from property lines, roads, and other features as follows:

- Distance from property lines of adjacent landowners without signed agreements with the Certificate Holder: 541 feet.
- Distance from property lines of landowners with signed agreements with the Certificate Holder: N/A.
- Distance from Bonneville/PSE transmission lines: maximum turbine height or as dictated by the respective transmission line owners.
- Distance from county and state roads: maximum turbine height.
- As noted above, distance shall be measured horizontally from the centerline of the turbine tower to the property line or the outermost edge of the road or other feature considered.

Although the Governor stated that "local sentiment about this project is just as important as it was for the Wild Horse Project," she also considered that the project could further statewide energy policy goals as stated in the Initiative 937, and said that she was "disappointed that the parties were unable to resolve the setback issue. The County did not articulate a well-reasoned setback standard during the hearing process." In this regard, her decision favored EFSEC: "As a result, EFSEC was left to articulate a standard for this project on a land use decision that should have had more local input. After careful review of the record, I commend EFSEC for its thorough, thoughtful review of this issue and indeed the entire project" (Gregoire 2007b: 2). After that, *The Seattle Times* published the two editorials excerpted below (Seattle Times 2007b; Riley 2007):

"Doubtful Double-back on State Wind Power," ***The Seattle Times*** **(2007b)**

Gov. Chris Gregoire's decision to ask a state panel to reconsider a controversial Kittitas County wind-farm permit seems disappointingly timid in comparison with her aggressive climate-change initiative.

That project's capacity already was reduced 10 percent from that of the initial project proposed five years ago. Why? Because of concessions to reduce the visual impact of the machines that stand more than a football field high. The Energy Facility Site Evaluation Council (EFSEC) permit includes reasonable setbacks and a requirement to turn off the turbines in some conditions.

The Kittitas Valley Wind Power Project has been vigorously opposed by landowners attracted to the bucolic area and by the Kittitas County commissioners.

The urgency of these changes suggests this is not a time to be appeasing people irritated about a view they don't own. The look of wind farms is a reason not to want them; it is no reason to stop them.

"Wind-farm Storm Shouldn't Blow Governor Off Course," ***The Seattle Times*** **(Riley 2007)**

Spurred by the urgency of climate change, Gregoire [Governor] is adamant about aggressively reducing Washington's greenhouse-gas emissions. She signed an executive order with ambitious goals and a multistate pact with fellow governors. The Legislature embedded her goals in law. Last fall, voters approved an initiative requiring utilities to get more of their power from non-hydropower renewable energy, such as this wind farm.

This latest dustup is only the beginning of what are bound to be many conflicts between rural lifestyles and aggressive new laws that nudge power companies to seek more and more wind energy.

At the council's hearing, neighbor squared off against neighbor. Some people don't want to look out the windows of their homes — or, in several cases, vacation homes — at the wind turbines that stand about two-thirds as tall as the Space Needle. They say proposed setbacks of about 1,600 feet are inadequate. Others don't mind the turbines' looks, are eager to collect the rent from leasing the ground under them and are enamored of their clean energy.

The council is expected to respond to the governor's request [to review the Site Agreement – letter after Order 826] in early August. It may stand by its initial recommendation or suggest the process be reopened to take new testimony. The council also might consider modifying its recommendation to give regulators more say over the siting of individual turbines as the project is under way. Expect a turbine-by-turbine battle, if that happens.

The state requires more wind power, not less. The governor should not abide too much more tinkering on the Kittitas Valley project just to save the views.

4.10 Reopening of the Adjudicative Proceedings and Final Approval

The Council determined that it would best be able to determine whether to reopen the adjudicative proceeding if all parties involved in the first proceeding and members of the public had an opportunity to comment on the governor's remand letter. For this reason, it held two public meetings on July 17, 2007 in Ellensburg, WA.

All parties that participated in the original adjudication attended the afternoon meeting, except for the Renewable Northwest Project (RNP), which provided written comments. Present in the meeting were the following: Sagebrush, represented by the attorneys Timothy L. McMahan and Erin L. Anderson; Counsel for the Environment, represented by Michael Tribble; Kittitas County, represented by Neil A. Caulkins; the Residents Opposed to Kittitas Turbines (ROKT), represented by James C. Carmody; F. Steven Lathrop, represented by the attorneys of law Jeff Slothower and F. Steven Lathrop; the Economic Development Group, represented by Debbie Strand; and Community Trade and Economic Development, represented by Tony Usibelli (Energy Facility Site Evaluation Council 2007i).

The evening meeting counted the presence of the Washington State Department of Community Trade and Economic Development, represented by Dick Fryhling; Department of Ecology, represented by Hedia Adelsman; Department of Fish and Wildlife, represented by Chris Towne; and Department of Natural Resources, represented by Judy Wilson, as well as Kittitas County, represented by Patti Johnson (Energy Facility Site Evaluation Council 2007j). According to Order 831 (Energy Facility Site Evaluation Council 2007d, p. 3), "39 members of the public addressed the Council. At least four of the sixteen non-participating landowners with residences directly affected by the Council's four times height setback standard provided their comments to the Council (three in person and one in writing). In addition, several other non-participating landowners with undeveloped land adjacent to the project addressed the Council."

During the meetings, the parties discussed whether additional opportunities for mitigation beyond the four times height standard could be identified (Energy Facility

Site Evaluation Council 2007d). In regard to the Governor Gregoire's concern about economic viability, EFSEC, the parties present in the meetings, and the general public agreed that the responsibility for determining the economic viability of this privately financed for-profit project should be Sagebrush. EFSEC, as a siting agency, stated that its focus is on the environmental and land use aspects, and that Order 826 had intentionally avoided economic viability justifications (Energy Facility Site Evaluation Council 2007d).

At the public meeting, Sagebrush was required to consider micro-siting to maximize the distance between turbines and the residences of non-participating landowners. Judge Torem raised the issue by saying (Energy Facility Site Evaluation Council 2007i, p. 40):

> It may be that the Governor is asking us to show (if there are) one or two particular turbines that need to be lopped off after further consideration? Council has to wrestle with that same factor that you are addressing today. But if the Applicant can come to us. I'm not asking for a sacrificial turbine. I'm asking simply for an approach in the micro-siting and language that would be binding on the applicant given further discretion to EFSEC that when it is weighing the factors of micro-siting that EFSEC, the on-the-ground manager, be able to prioritize that particular visual setback, which apparently caught the Governor's eye.

Mr. McMahan, representing Sagebrush, agreed to do the micro-siting proposal and stated that "if we can steer this in the direction of micro-siting, it does provide us an opportunity in engineering the project and working with EFSEC staff to maintain as many as the Applicant can while addressing a question that the Governor has asked us to address" (Energy Facility Site Evaluation Council 2007i, p. 43).

It was finally concluded that micro-siting was the only economically viable method to achieve the additional setbacks requested by the Governor. After the meetings, EFSEC issued Order 831, in which it stated that additional provisions should be added to the certification agreement Article I (C)(7) regarding micro-siting of turbines (Energy Facility Site Evaluation Council 2007d, p. 4). EFSEC Chair Jim Luce said "the Council heard the Governor loud and clear and has taken additional steps to help minimize potential visual impacts raised by some non-participating landowners in the affected area." With this additional mitigation measure, in August of 2007, EFSEC recommended again that the governor approve KVWPP. Their press release stated that "The permit modification provides that in the final siting of turbines within 2500 feet of a non-participating landowner's existing residence, the project developer shall give 'highest priority' to increasing the distance of the turbine from that non-participating landowner's residence, even beyond the minimum four times height setback" (Energy Facility Site Evaluation Council 2007f, p. 2).

On September 18, 2007, the governor finally approved the project as amended by Order 831. In her acceptance letter, she re-stated her commitment to renewable energy: "our legislators and citizens have recently articulated their strong preferences for renewable resources. Those policies are not in doubt and I remain committed to them. To fulfill them we will have to build infrastructure that broadly benefits our citizens, and may impose burden on some" (Gregoire 2007a, p. 1). She also recognized the impasse created with the "local government" (Gregoire 2007a, p. 1–2):

> I have given careful consideration to this decision and do not exercise my authority lightly. As Governor, I have seen the federal government take and exercise authority that I would have preferred to keep in the state's control, and so I know precisely how local government perceives it when the state does the same thing. Nonetheless, the Legislature has given the Governor the authority to act to ensure that all citizens of the state continue to have access to abundant, affordable, and environmentally preferred power. I find the underlying policy rationale and its benefits compelling in my decision to move forward.

4.11 Opposition to the Governor's Decision and Petition to the Supreme Court

Not all observers understood the governor's letter as representing her good intention and compliance with Washington State's goals. In an editorial for *The Seattle Times* entitled "Decision on Wind Farm sets a Dangerous Precedent," guest columnist Jeff Howard argued as follows (Howard 2007):

> GOV. Christine Gregoire just allowed an unelected commission known as the Energy Facility Site Evaluation Council (EFSEC) to preempt a local land-use decision by the Kittitas County Board of Commissioners. The issue of siting huge wind-turbine installations has been a bit under the radar screen in Western Washington, but the governor's action should raise alarms all over the state. If this massive overreach of state authority is not reversed, we can be certain that the EFSEC process will be used by companies to allow more and bigger wind installations anywhere in our state.

Evidence of this concern, according to Howard, was that, within hours of the governor's decision, enXco petitioned EFSEC to reopen an application for a wind-turbine installation that had previously been denied by Kittitas County. The columnist said that Kittitas County officials acted with "good faith and careful decision-making when they earlier approved the Wild Horse Wind Farm project in an appropriate location in the eastern part of the county (Howard 2007)." The western Kittitas Valley was also described as "one of the most pristine and treasured parts of our state" (Howard 2007). Howard concluded, "There will be absolutely no protection for anyone living anywhere. Great swaths of Washington will be festooned with giant turbines, which create very little energy but destroy property values and quality of life for miles in all directions" (Howard 2007).

Petitions for Judicial Review of both the governor's approval and EFSEC's recommendation were filed with the Thurston County Superior Court in October of 2007, by Kittitas County, ROKC, and Steven Lathrop. On February 29, 2008, the Thurston County Superior Court certified the case record and transferred it to the Washington Supreme Court. The petition claimed that the county, not the state, holds preeminent authority for reviewing and permitting power plants. The second main argument was that EFSEC's process was biased, particularly on the part of EFSEC Chair Luce (Wind Works Northwest 2008).

Even though there was much resistance to KVWPP, a survey of 300 Kittitas County voters conducted by a Portland-based public-opinion research firm (Moore Information) some months prior to this court process demonstrated that 78% of Kittitas County voters supported the development of wind farms within the county (Electric Energy Online 2008).

4.12 State Supreme Court's Decision

Acknowledging the size of the 400-foot wind towers, the court quoted from the literary classic *Don Quixote De La Mancha*, by Miguel De Cervantes: "They are giants: and, if thou art afraid, get thee aside and pray, whilst I engage with them in fierce and unequal combat" (Woodward 2008). Issues of project siting were considered by the court; the reality however, was that no matter how sympathetic the judge may have been to the residents' concerns, they were not the legal issue in the case. James Carmody, the local attorney who represented ROKT, stated: "When you get to the supreme court, what you're dealing with are legal issues, not fact-specific siting decisions […] So the real question was, who gets to make the decision in the state of Washington as to a wind power facility or an alternative energy facility?" (Wind Power Monthly 2009).

On this question of authority for siting approval, EFSEC Chair Luce said that "if EFSEC can't preempt [the county]… it will lose any credibility as a 'state siting council.' The signal to counties opposed to siting energy projects of any sort will be clear: devise an unworkable process with no rules which duplicates EFSEC's process… and you may not have to site energy projects" (Wind Works Northwest 2008). On the other hand, a decision that the state could overrule the county's authority would raise other concerns: other companies with wind-farms projects would go straight to EFSEC as a way to override local concerns (Wind Power Monthly 2009).

In November of 2008, the State Supreme Court upheld the governor's decision and EFSEC's project approval. Gabriel Alonso, Chief Operating Officer for Horizon Wind Energy, said that this "unanimous Supreme Court ruling is significant for Horizon Wind Energy and for Washington citizens. The Court, in upholding the authority of the Energy Facility Site Evaluation Council (EFSEC), affirms that much-needed power projects that pass the State's rigorous review will be built and that Washington State's goal of developing new renewable energy can truly be achieved" (Horizon Wind Energy 2008).

The Court decided that in cases where local jurisdiction policies duplicate state processes, the state (through EFSEC) has the authority to preempt. This decision corroborates the purpose of EFSEC, which was to be a "one stop" permitting agency that could make decisions based on the public interest. Regarding the allegation of bad faith on the part of the applicant and EFSEC, Judge Richard Hicks said: "I don't think there has been a minimum threshold here that substantiates the allegation of impropriety" (Wind Works Northwest 2008).

McMahan and Anderson, two of the attorneys who led the permitting effort and subsequent litigation, wrote the following on the Stoel Rives Attorneys website (McMahan and Anderson 2008):

> In upholding the state's authority to approve energy facility siting, the Court confirmed that neither the statewide land use planning law nor local land use laws adopted pursuant to the Growth Management Act supersede or repeal EFSEC's preemption powers. This holding is highly significant for energy facility policy and planning, as it ensures a coordinated, expedited process intended to ensure abundant and affordable power, better enabling the state to achieve its goals to reduce CO_2 emissions and implement its renewable portfolio standard.

In addition, they affirmed that this case was particularly important because it "had been closely watched not only by wind industry stakeholders but by the broader energy industry as well, as the challenge to the Energy Facilities Site Locations Act (EFSLA), ch. 80.50 RCW, sought to invalidate the entire statutory siting framework and remove renewable energy facilities from EFSEC's jurisdiction" (McMahan and Anderson 2008).

4.13 Amendment to the Site Certification Agreement

Sagebrush, in a letter dated May 29, 2009, requested that EFSEC amend the Site Certification Agreement: "The primary purpose of this amendment is to reduce the number of wind turbines from 65 to 52, along with associated reduction and minimization of environmental impacts, including the impacts on non-participating residences, while maintaining the full power generation of the Project" (Energy Facility Site Evaluation Council 2009: 1). To achieve this, KVWPP proposed the following Amendment to the Site Certification Agreement (Energy Facility Site Evaluation Council 2009, p. 1):

1. Reduce the number of turbines allowed, eliminating certain wind turbine locations and all or portions of certain corridors.
2. Adjust certain wind turbine approximate locations and corridors.
3. Relocate the Operations and Maintenance facility to a site further away and less visible from Hwy. 97.
4. Add a 6.29-acre parcel to reduce ground disturbance related to collection line routing under existing roads.
5. Reduce the total project area to approximately 5416 acres, reducing the temporary disturbance of ground area by 14.4 acres, and reducing permanent disturbance by 13.3 acres.

This request was a result of Sagebrush's work in anticipation of the micro-siting process. The micro-siting was as a setback condition established to respond to the Governor's demand (see Resolution 831). The setbacks proposed by Sagebrush in the amendment were possible due to their access to higher-capacity wind turbine generators that were not previously available. This access to new technology, in addition to adjustments in the approximated position of turbines, would allow them to reduce the number of turbines and corridors near residences. The most significant changes in terms of distance and density occurred in one specific area. There, six towers were removed and several others were moved. Construction was expected to commence in the fall of 2009 and to be completed in 2010 (Energy Facility Site Evaluation Council 2009).

A hearing was held in Ellensburg on June 23, 2009. Notice about the hearing was sent to about 380 people. Written and oral comments were received from affected residents Wayne Bell and Michael Robertson regarding the proposed placement of two wind turbines (A-1 and A-2) that would be closer than in the original hypothetical layouts, as approved within the micro-siting corridors. They also mentioned that these two turbines would cause "health" effects related to "shadow flicker" due to the kind of equipment that Sagebrush wanted to use (Energy Facility Site Evaluation Council 2009: 4). An agreement was made to move the turbines in question and to avoid running the turbines when shadow flicker was possible (Energy Facility Site Evaluation Council 2009: 8).

Regarding the siting of turbines, EFSEC concluded that, overall, both residents were better-off because several turbines located near the 2500 feet threshold were being removed. "In the totality of the circumstances, the Council finds that the Applicant has satisfied the spirit and intent of the setback condition as it relates to the Robertson residence. The Council therefore finds that the Applicant has complied with the setback condition with respect to these turbine locations" (Energy Facility Site Evaluation Council 2009, p. 9).

4.14 KVWPP Implementation and Aftermath

KVWPP construction finally started in 2010. "We had hoped to have a 2009 construction schedule, but right now the plans are to start turbine unit installations in spring 2010," said Joy Potter, Horizon's Kittitas Valley project manager (Johnston 2009). Forty-eight turbines were installed by 2010, each with a tip height of 406 feet, spread out on 5416 acres. The farm was fully operational and generating power into the electrical grid on December 15, 2010.

In 2011, KVWPP Operations Manager Eric Melbardis noted: "People have asked why do they see the turbines at our project stopped sometimes, but they look farther west and see the turbines at the Wild Horse project spinning" (Johnston 2011). This happens because KVWPP did not have a long-term power purchase agreement with any utility company to buy all the electricity generated by the project. In 2011, their sales were dependent upon the right price offers and wind speed that would make the transaction economically feasible. "If the wind speed is not adequate and the price buyers are willing to pay is not adequate, the blades don't turn" (Johnston 2011).

Other projects in the county, such as Wild Horse Wind and Solar Facility and Vantage Wind Power Project, already had agreements to sell their energy to a western Washington system and to a San Francisco-based utility. Manager Melbardis did not think KVWPP was in a bad position, however: "It is not unusual to have a new (wind farm) site sell on the daily market before a long-term purchase agreement can be developed for its power output" (Johnston 2011). In February of 2012, in an EFESC Meeting, he gave the statement below (Energy Facility Site Evaluation Council 2012, p. 9):

> For our project status for the month of January, Kittitas Valley Wind Power Project produced 13,055 megawatt-hours with average wind speeds of 4.7 meters a second, which is 10.8 miles per hour.
>
> We have no safety incidents to report. The project has continued to operate without lost time or OSHA-recordable incidents since its inception. The project is in compliance as of February 12, 2012. We have no new sound complaints. As Mr. Wright explained, we do have a meeting today to discuss the Robertson complaint.
>
> We did submit our Operations Phase Sound Study to EFSEC on December 21, 2011. The project is in compliance with the SCA and WAC. Our nightly hourly level is below 50 decibels. Shadow flicker. We have no complaints there. Our automatic curtailment is still working as required.
>
> For environmental, we have no environmental incidents to report. Weed and re-vegetation surveys will continue in the spring. Avian and bat fatality monitoring, as I explained in the past, is a year-round study. That is continuing. No storm water discharge to report.
>
> We finalized the TAC meeting minutes from our December TAC meeting, and it has been provided to the EFSEC Staff and are available for Council review as needed. And we are continuing to monitor all stormwater BMPs, improving them as necessary.

4.15 Lessons Learned

From an environmental and energy development tradeoff perspective, some may feel that the societal benefits of rapidly expanding renewable energy may justify circumvention of public participation. In the case of the Kittitas Valley Wind Power Project, however, former Washington Governor Christine Gregoire set a precedent for top-down action when she cited Washington State's RPS to justify approving KVWPP after the local county government had already voted against the development.

The current climate crisis may present a scenario where swift and centralized administrative decision-making is preferable from a macro perspective. Nonetheless, one still can question whether the collective benefits of renewable energy development justify contradicting the goal to "encourage public involvement in decisions" outlined in the Washington State and national environmental protection acts (Washington Administrative Code [WAC] 197-11-030). Although once simplistically labeled as NIMBY, research has since highlighted the important role of local opposition to "improve the environmental and social justice credentials of proposals" and to produce a more "nuanced evaluation" of project impacts (Mcclymont and O'hare 2008).

References

Cappiello, D. (2013, November 22). Wind energy company pleads guilty to eagle deaths. *The San Diego Tribune*. http://www.sandiegouniontribune.com/sdut-wind-energy-company-pleads--guilty-to-eagle-deaths-2013nov22-story.html. Accessed 15 Aug 2016.

Cornwall, W. (2007a, June 26). An ugly battle over clean power. *Seattle Times*. http://www.seattle-times.com/html/localnews/2003762626_wind26m.html

Cornwall, W. (2007b, September 19). Gregoire approves wind-power project. *Seattle Times*. http://www.seattletimes.com/seattle-news/gregoire-approves-wind-power-project/

Electric Energy Online. (2008, April 11). Desert claim wind power will ask EFSEC to pre-empt Kittitas County and recommend approval of wind farm to power nearly 60,000 homes. *Electric Energy Online*. http://www.electricenergyonline.com/?page=show_news&id=65755

Energy Facility Site Evaluation Council. (2004). Applicant's request for preemption. http://www.efsec.wa.gov/kittitaswind/preemption/Preemption%20Request.pdf Accessed 15 Aug 2016.

Energy Facility Site Evaluation Council. (2006). Applicant's second request for preemption. http://www.efsec.wa.gov/kittitaswind/preemption/Exhibits%201/KV%20Second%20Request%20for%20Preemption.pdf. Accessed 15 Aug 2016.

Energy Facility Site Evaluation Council. (2007a). *Chapter 1: Summary*. Olympia: Washington State Energy Facility Site Evaluation Council. http://www.efsec.wa.gov/Tesoro%20Savage/SEPA%20-%20DEIS/DEIS%20PAGE.shtml. Accessed 15 Aug 2016.

Energy Facility Site Evaluation Council. (2007b). *Council Order 826: Findings of fact, conclusions of law, and order recommending approval of site certification on condition*. Olympia, WA: Washington State Energy Facility Site Evaluation Council. http://www.efsec.wa.gov/kittitaswind/Recommendation%20to%20Governor%205-2-07/Letter%20to%20Governor%20stamped%20received%205-2-07.pdf. Accessed 15 Aug 2016.

Energy Facility Site Evaluation Council. (2007c). *Council Order 827: Order on motions for reconsideration of order 826 by Kittitas County, and Mr. Ed Garrett and Ms. Rosemary Monaghan*. Olympia: Washington State Energy Facility Site Evaluation Council. http://www.efsec.wa.gov/FILES/orders/827.pdf. Accessed 15 Aug 2016.

Energy Facility Site Evaluation Council. (2007d). *Council order 831: Order on remand letter from Governor Gregoire modifying the draft site certification agreement accompanying Order No. 826*. Olympia: Washington State Energy Facility Site Evaluation Council. http://www.efsec.wa.gov/kittitaswind.shtml. Accessed 15 Aug 2016.

Energy Facility Site Evaluation Council. (2007e). *EFSEC recommends governor approve Kittitas Valley wind power project*. Olympia: Washington State Energy Facility Site Evaluation Council. http://www.efsec.wa.gov/kittitaswind.shtml. Accessed 15 Aug 2016.

Energy Facility Site Evaluation Council. (2007f). *Fact sheet*. Olympia: Washington State Energy Facility Site Evaluation Council. http://www.efsec.wa.gov/kittitaswind.shtml. Accessed 15 Aug 2016.

Energy Facility Site Evaluation Council. (2007g). *Introduction*. Olympia: Washington State Energy Facility Site Evaluation Council. http://www.efsec.wa.gov/kittitaswind.shtml. Accessed 15 Aug 2016.

Energy Facility Site Evaluation Council. (2007h). *Site certification agreement between the State of Washington and Sagebrush power partners, LLC, for the Kittitas Valley wind power project*. Olympia: Washington State Energy Facility Site Evaluation Council. http://www.efsec.wa.gov/kittitaswind.shtml. Accessed 15 Aug 2016.

Energy Facility Site Evaluation Council. (2007i) *Special afternoon meeting*. Olympia: Washington State Energy Facility Site Evaluation Council. http://www.efsec.wa.gov/kittitaswind/Remand%207-07/Order%20831.pdf. Accessed 15 Aug 2016.

Energy Facility Site Evaluation Council. (2007j). *Special evening meeting*. Olympia: Washington State Energy Facility Site Evaluation Council. http://www.efsec.wa.gov/default.shtm. Accessed 15 Aug 2016.

Energy Facility Site Evaluation Council. (2008). *EFSEC recommends the governor approve the Kittitas Valley wind power project*. Olympia: Washington State Energy Facility Site Evaluation Council. http://www.efsec.wa.gov/kittitaswind.shtml. Accessed 15 Aug 2016.

Energy Facility Site Evaluation Council. (2009). *Resolution 328*. Olympia: Washington State Energy Facility Site Evaluation Council. http://www.efsec.wa.gov/kittitaswind/09%20Amend/Res%20328.pdf. Accessed 15 Aug 2016.

Energy Facility Site Evaluation Council. (2012). *Monthly council meeting*. Olympia: Washington State Energy Facility Site Evaluation Council. http://www.efsec.wa.gov/FILES/minutes/2012/5-22-12%20Council%20Meeting%20Minutes.pdf. Accessed 15 Aug 2016.

Energy Facility Site Evaluation Council. (2013). *Homepage*. Olympia: Washington State Energy Facility Site Evaluation Council. http://www.efsec.wa.gov/default.shtm. Accessed 15 Aug 2016.

Garret, E., & Monaghan, R. (2007, April 6). *Reconsideration document for KVWPP*. http://www.efsec.wa.gov/kittitaswind.shtml. Accessed 15 Aug 2016.

Gregoire, C.O. (2007a). *Response to Kittitas Valley wind power project EFSEC recommendation letter.* http://www.efsec.wa.gov/kittitaswind.shtml. Accessed 15 Aug 2016.

Gregoire, C.O. (2007b, August 14). *Kittitas Valley wind power project, EFSEC recommendation letter*. http://www.efsec.wa.gov/kittitaswind.shtml. Accessed 15 Aug 2016.

Horizon Wind Energy. (2008, November 20). Washington State Supreme Court upholds EFSEC permit for Kittitas Valley wind power project. *PR Newswire*. http://www.prnewswire.com/news-releases/washington-state-supreme-court-upholds-efsec-permit-for-kittitas-valley-wind--power-project-65496647.html. Accessed 15 Aug 2016.

Howard, J. (2007, October 11). Decision on wind farm sets a dangerous precedent. *Seattle Times*. http://www.seattletimes.com/html/opinion/2003940563_windfarm11.html

Johnston, M. (2007, May 4). Gregoire to decide on Kittitas Valley wind farm project. *The Wenatchee World*. http://www.wenatcheeworld.com/news/2007/may/04/gregoire-to-decide-on-kittitas-valley-wind-farm/. Accessed 15 Aug 2016.

Johnston, M. (2009, January 26). No KVWPP wind farm this year. *Daily Record*. http://www.kvnews.com/articles/2009/01/24/news/doc497a64257246c260169237.txt. Accessed 15 Aug 2016.

Johnston, M. (2011, April 6). At Kittitas Valley project, turbines don't always turn. *Daily Record*. http://www.dailyrecordnews.com/news/at-kittitas-valley-project-turbines-don-t-always-turn/article_1342474-6077-11e0-9730-001cc4c002e0.html. Accessed 15 Aug 2016.

Kittitas County Board of County Commissioners. (2006). *Resolution 2006–90*. County of Kittitas.

Kittitas County Board of County Commissioners. (2007). *Kittitas County's motion for reconsideration*. County of Kittitas.

Mcclymont, K., & O'Hare, P. (2008). We're not NIMBY's! Contrasting local protest groups with idealized conceptions of sustainable communities. *Local Environment, 13*(4), 321–335.

McMahan, T.L., & Anderson, E.L. (2008, December 15). *Renewable energy facility siting legislation on the right track*. Stole & Rives, LLP. http://www.stoel.com/showarticle.aspx?Show=3454. Accessed 15 Aug 2016.

News Tribune (2007). State energy battles are blowing in the wind. *The News Tribune*. http://www.thenewstribune.com/2007/04/10/v-lite/35154/state-energy-battles-are-blowing.html. Accessed 15 Aug 2016.

Ohrenschall, M. (2003, July 16). *Third major wind farm proposed for central Washington*. http://www.bluefish.org/3rdwind.htm. Accessed 15 Aug 2016.

Pacific Power. (2013). *Kittitas Valley wind farm*. http://www.pacificpower.net/env/bsre/bses/kvwf.html. Accessed 15 Aug 2016.

Riley, K. (2007, July 23). Wind-farm storm shouldn't blow governor off course. *Seattle Times*. http://www.seattletimes.com/html/opinion/2003799034_riley23.html. Accessed 15 Aug 2016.

Sands, C., & Sands, T. (2007). *Please deny KVWPP and desert claim wind power project*. http://www.thepetitionsite.com/takeaction/896/779/246/. Accessed 15 Aug 2016.

Schantz, L. (2007). Please deny KVWPP and desert claim wind power project. CTC Farm. www.ctcfarm.com/petition.html. Accessed 15 Aug 2016.

Seattle Times. (2007a, April 10). Editorial: An inconvenient view. *The Seattle Times*, April 10. http://www.seattletimes.com/html/editorialsopinion/2003659455_horizoned10.html

Seattle Times. (2007b, July 5). Editorial: Doubtful double-back on state wind power. *The Seattle Times*. www.seattletimes.com/html/editorialsopinion/2003773636_winded05.html

Shapiro and Associates. (2003). Kittitas Valley wind power project scoping summary. http://www.efsec.wa.gov/kittitaswind/Scoping%20Summary%20web.pdf. Accessed 15 Aug 2016.

Suzlon. (2015). *S88 wind turbine brochure*. http://www.suzlon.com/pdf/S88%20product%20brochure.pdf. Accessed 15 Aug 2016.

U.S. Department of the Interior. (2016). "Wind energy basics" wind energy development programmatic EIS information center. http://windeis.anl.gov/index.cfm

Wind Power Monthly. (2009, January 1). Local opposition put in its place - fundamental court ruling. *Wind Power Monthly*. http://www.windpowermonthly.com/article/953391/local-opposition-put-its-place---fundamental-court-ruling. Accessed 15 Aug 2016.

Wind Works Northwest. (2008). *Kittitas County and its NIMBY allies lose big to usurp Washington State's authority to permit power plants*. http://www.wind-works.org/cms/index.php?id=85&tx_ttnews%5Btt_news%5D=263&cHash=16ba9e9c55da86a17380aa8d6cb740ef. Accessed 15 Aug 2016.

Woodward, C. (2008, November 20). Supreme Court upholds Kittitas County wind farm. *Seattle Times*. http://www.seattletimes.com/html/localnews/2008415591_webwindfarm20m.hhtm

Yakima Herald-Republic. (2007, April 3). Governor should OK revised wind farm. *The Yakima Herald-Republic*. http://www.highbeam.com/doc/IP2-9640804.html. Accessed 15 Aug 2016.

Zilkha Renewable Energy. (2003, January 13). *Cover Letter*. http://www.efsec.wa.gov/kittitaswind/cover%20letter%20final.pdf. Accessed 15 Aug 2016.

Chapter 5
Newberry Crater Geothermal

Abstract This case study considers the development of geothermal power near Oregon's Newberry Volcano in central Oregon. The Seattle-based AltaRock Energy Company began cracking hot rock deep within Newberry Volcano in 2010 by pouring water down a 10,000-foot well. Work was then suspended in 2012 due to weather conditions and the need to test and reinforce a concrete casing extended 5300 feet into the well. While there is more testing planned and the company needs to identify additional funding, the plan is for the future development of the resource with full production still years away. However, this nascent technology has created some controversy. Critics are concerned about the possibility of earthquakes and potential pollution of the aquifer, but supporters claim neither issue has been documented to date. In this context, this case study outlines the developing debate between the need to develop new non-carbon based energy sources versus potential environmental tradeoffs.

Keywords Geothermal power • Newberry Volcano • Earthquakes • Pollution

5.1 Introduction

> *I'm not against geothermal energy, it's just not a good idea to put (a power plant) in the middle of a national monument.* (Bill Marlett, Executive Director of the Oregon Natural Desert Association (Friedrichs 2007))
>
> *We're trying to get the science and figure out if this is a viable alternative energy source...Maybe it won't be, but they need to look at all avenues.* (Linda Christian, Environmental Coordinator for the BLM's Prineville District, Oregon)

Geothermal energy is seen as an important potential source of clean energy because it is said to have a small environmental footprint, emits few greenhouse gases, and can be produced consistently and continuously (Petty 2010). A report titled *The Future of Geothermal Energy* produced by the Massachusetts Institute of Technology (2006) investigated the potential of geothermal energy in the United States. It found that geothermal energy could provide 100,000 MW or more in the next half century through the Enhanced Geothermal Systems (EGS) technology (Massachusetts Institute of Technology 2006).

J.C. Pierce, B.S. Steel, *Prospects for Alternative Energy Development in the U.S. West*, Environmental Challenges and Solutions 4,
DOI 10.1007/978-3-319-53414-5_5

Fig. 5.1 Example of Geothermal Energy Facility (National Renewable Energy Laboratory)

To access this source of energy, a very deep well of approximately 10,000 feet may be used to create an engineered geothermal system (also known as an "Enhanced Geothermal System" or "EGS"). Cold water is then injected into a network of small cracks/fractures in an area with geothermal activity, such as near an ancient volcano. As it passes over the surrounding hot rock, the water is heated to temperatures high enough for geothermal energy production upon its return to the surface in the form of steam. The steam is then used to turn turbines for energy production. Once this heated fluid is used for power production, it is re-injected into the reservoir of small cracks to be reheated, thus creating a relatively continuous heating/power production cycle (Simon 2007, pp. 123–144). An example of a geothermal facility is shown in Fig. 5.1.

5.2 Case Study Project

The Newberry Volcano is a large shield volcano located about 20 miles southeast of Bend, Oregon, and about 40 miles east of the Cascade Range. There are two lakes in its caldera (Paulina and East lakes). As early as 1981, scientists found that temperatures in the Newberry Caldera can reach upwards of 540 °F at 3000 feet below the floor. Thus, the volcano has significant geothermal potential: not only is its temperature the highest ever recorded at a dormant volcano, but it is also hotter than the geysers of California—the largest producer of geothermal energy in the world.

In 1990, Congress issued the Monument Act and designated over 50,000 acres to become a national monument through the Newberry National Volcanic Monument (NNVM) Act. The act restricts geothermal development within the NNVM and provides specific guidance to activities outside of monument boundaries. The Monument Act also includes a provision for the maintenance of a research and monitoring program for geothermal resources in the vicinity of NNVM and Special Management Area (Bureau of Land Management 2011b). According to the Public Law 101–522, Section 8(a):

> Nothing in this Act shall be construed as authorizing or directing the establishment of protective perimeters or buffer zones around the Monument or Special Management Area for the purpose of precluding activities outside the Monument and Special Management Area boundary which would otherwise be permitted under applicable law [...].The fact that activities or uses outside the Monument and Special Management Area can be seen, heard, measured, or otherwise perceived from within the Monument and Special Management Area shall not, of themselves, limit, restrict, or preclude such activities or uses up to the boundary of the Monument and the Special Management Area" (Bureau of Land Management 2011b, p. 20).

As mandated in the Monument Act, the USDA Forest Service prepared a Comprehensive Management Plan (CMP) for the NNVM, establishing programmatic management direction for National Forest lands within the newly created monument and defining valid geothermal leases within the Special Areas identified within the Monument Act:

> Leases located outside the Monument boundary were designated for geothermal use by a grassroots citizens' committee that included local representatives of the community, environmental groups, government and the geothermal industry. This Monument Committee created the vision and helped draft the legislation that was adopted in the Congressional process that established the Newberry National Volcanic Monument and adjacent geothermal leases (AltaRock Energy/Davenport Newberry 2010).

The Newberry geothermal project is located in an area of the Deschutes National Forest identified as appropriate for future geothermal exploration in the 1990 Deschutes National Forest Land and Resource Management Plan (LRMP), as amended. This area was also recognized as being appropriate for geothermal use in the subsequent federal legislation that created the NNVM (Public Law 101–522) (Bureau of Land Management 2011b).

5.3 Project Partnerships

The Newberry Crater Geothermal Project involves collaboration of the AltaRock Energy and Davenport Power partnership, as well as with students from the University of Oregon, Lawrence Berkeley National Laboratory, University of Utah, Texas A&M, Temple University, and scientists from the U.S. Geological Survey (Petty 2010). The first phase of the project (planning for well stimulation, environmental studies, and public outreach) also included the participation of the State of Oregon Departments of Geology and Mineral Industries, Environmental Quality, and Water Resources; Lawrence Livermore Laboratory; Los Alamos Laboratory; Sandia National Laboratory; and Oregon State University (Newberry Demonstration 2012).

The Newberry Crater Geothermal Project was subjected to regulatory approval and oversight from several agencies, including the USDI Bureau of Land Management (BLM), the Forest Service, the U.S. Department of Energy (DOE), and several Oregon agencies (Petty 2010). Doug Perry, President of Davenport Newberry, said in a June 8, 2010, press release:

> We see this project as a true public-private collaboration, and we welcome input from government officials and community members. We seek nothing less than an ongoing 360-degree conversation with all parties. In addition, we've studied the science, refined the technology, and listened attentively to the community; we've also tried to use and deploy thoughtful planning and control systems" (AltaRock Energy/Davenport Newberry 2010).

5.4 Exploratory Well Drilling

According to a Newberry project's press release, "for decades, Newberry Volcano has been considered as high potential for commercial geothermal resource. The difficulty is in finding those specific locations deep underground where the resource can be accessed for long-term production. A successful exploration program is not guaranteed to be quick, or easy" (Davenport Power 2008). After extensive geophysical and EGS studies were conducted in 2006, Davenport was finally ready to drill exploratory test wells in order to explore the potential geothermal resources of the Newberry area; it sought underground geologic features that had the right combination of heat and water to produce steam (Altarock/Davenport Power 2007). In a press release, Davenport stated that it "is committed to developing, constructing and operating this project in a safe, environmentally responsible manner. We continue to be ready and willing to work with all of the stakeholders and each of the regulatory agencies to ensure a quality Project" (Altarock/Davenport Power 2007).

After working with the BLM and the Forest Service, among others, Davenport was able to successfully complete an environmental analysis, and, on November 26, 2007, they gained the permit to proceed with road maintenance and pad preparation for their exploratory well drilling program. Davenport obtained environmental approval from the BLM to prepare three well pads of approximately five acres each. This preparation involved clearing trees from the sites, along with grading and compacting the ground. According to Davenport, "two of the sites have been fully logged in the past and the third site contains a mix of new growth and some older trees. We are working on the specific placement of the third well pad site to leave as many of the older and larger trees as practical" (Altarock/Davenport Power 2007). The exploratory wells provided a better understanding of the Newberry geothermal resources; they were important to understanding whether favorable conditions were in place to develop a power plant facility. Well pads and wells were located on federal geothermal leases held by the project, located about 25 miles south of Bend and west of the Newberry National Volcanic Monument (ThinkGeoEnergy 2010).

Drilling for the first exploratory well started in April of 2008. Drilling of Exploration Well #1 was completed on July 12, 2008, at a depth of 10,060 feet. "While this well encountered high temperatures and some fractures, the underground rock formations did not contain significant quantities of water or steam. Davenport Power will continue to study this well and use the geologic and scientific data throughout the exploratory drilling program" (ThinkGeoEnergy 2010). Following completion of Exploration Well #1, Davenport started drilling the next exploration well approximately two miles north. Drilling of Exploration Well #2 began August 11, and was completed on October 19, 2008, at a depth of 11,599 feet (ThinkGeoEnergy 2010).

Even though both exploratory wells found high temperatures and some fractured rock, continued exploratory efforts were needed in order to locate geothermal resources sufficient to supply a power plant. In the winter of 2008, Davenport gathered scientists to start analyzing an extensive amount of scientific and geologic data in order to better understand the deep underground. Based on that assessment, Davenport Power believed it would be able to develop the best strategy for successful development of the next exploratory wells (ThinkGeoEnergy 2010).

5.5 Surface Geophysical Studies and Gradient Drilling

Until August of 2009, no other exploration wells were drilled; meanwhile, during this period, the Newberry Project team began to implement another exploration phase approved on BLM-leases. They planned surface geophysical work to collect data about the nature and geology of the Newberry Volcano. The surface work, planned to start in 2010, was not invasive and had low environmental impacts (ThinkGeoEnergy 2010).

In 2009, the Newberry Geothermal Project (NGP) also proposed to drill a series of small, narrow temperature gradient holes as much as 3500 feet deep in up to 12 locations outside the NNVM. These holes represented another method of exploration intended to further scientists' and geologists' understanding of the subsurface geology of the western flanks of the Newberry Volcano. In their proposal, the 12 sites would be approximately 100 feet by 100 feet in size, and would be placed along existing Forest Service roads at previously logged areas (ThinkGeoEnergy 2010).

This proposed temperature gradient drilling activity had to be approved by the BLM (Prineville Office) through an Environmental Assessment (EA), pursuant to the National Environmental Policy Act (NEPA). The BLM and the Forest Service (Rock Ranger District) conducted the assessment of these activities. The EA completion was expected by early 2010 and the field exploration on their leases at Newberry Volcano was expected to start by Spring 2010. As stated by Doug Perry, President of NGP: "We remain committed to exploring, developing and operating at Newberry Volcano in a safe manner, as good stewards of the public land and the geothermal resource" (ThinkGeoEnergy 2010).

5.6 Site Certification

In 2008, Newberry Geothermal Holdings submitted a Notice of Intent (NOI) to the Oregon Department of Energy in order to apply for a Site Certificate for the Newberry Geothermal Project (see Fig. 5.1). The site certificate process is a consolidated permitting process that includes the review of applicable regulations of most other state agencies. The applicant intended to satisfy the Council's Land Use Standard by seeking a Council determination under ORS 469.504(1) (b) instead of seeking a land use determination by Deschutes County. The Site Certificate process does not include the issuance of federal permits or permits that the federal government has delegated to other state agencies (such as air quality permits delegated to the Department of Environmental Quality). The NGP was reviewed by the Forest Service (because the project is located on federally managed land) and the BLM (which has jurisdiction over geothermal leases). The federal agency review is required under NEPA. According to the NOI for the project (Oregon Department of Energy 2008),

> The Newberry Geothermal Project would include power generation systems, geothermal fluid systems, transmission lines, access roads, operations and maintenance buildings and other related or supporting facilities. In addition, the Project would have up to four separate electric generating units that would produce 35 to 45 megawatts. Each generation turbine would consist of a turbine, generator, condenser, cooling water pumps, gas removal system, cooling tower and other equipment. Production and injection wells would be 6,000 to

> 10,000 feet deep and that their location would depend on efficient utilization of the geothermal resource while minimizing environmental impacts.
>
> Regarding transmission and contracts for the output of the project, a 230-kilovolt transmission line, approximately 12 miles in length, would carry power from the Newberry Geothermal Project to a Bonneville Power Administration (BPA) substation near La Pine. Davenport had executed transmission service agreements with BPA for long-term transmission of the project's power output to Pacific Gas & Electric Company. Davenport has a contract for the sale of 120 megawatts of NGP output for a period of 20 years beginning as early as 2010.
>
> Finally, the Newberry Geothermal Project would lie completely outside the boundaries of the Newberry National Volcanic Monument. The area under consideration is approximately 9,000 acres, but the project would occupy less than 5 percent (450 acres) of this total

On September 18, 2008, the Oregon Department of Energy held an early public information meeting in Bend. The purpose of the meeting was to provide the public with preliminary information about the proposed project and Oregon's Energy Facility Site Certificate process. According to a Newberry Geothermal Project press release, approximately 30 people attended the meeting, including citizens, agencies, and representatives from various organizations. On October 31, 2008, the Oregon Department of Energy held an Oregon Geothermal Working Group meeting at the Central Oregon Environmental Center in Bend. Part of this meeting was an environmental outreach session for the public and interested agencies in which representatives from geothermal projects around the state took the opportunity to speak about their projects, including representatives of NGP. According to another NGP press release, over 50 people attended, including conservation organizations, citizens, media representatives, and Deschutes County officials (ThinkGeoEnergy 2010).

5.7 Project Funding

After acquiring the permit for exploratory wells (as described in the section above), Davenport obtained the support of U.S. Renewables Group and Riverstone Holdings; both groups agreed to provide development capital for NGP, particularly to drill exploratory wells (Business Wire 2008). The following was said about this partnership (Business Wire 2008):

> Doug Perry, President of Davenport Power stated: "we are very pleased in the confidence expressed by this investment from US Renewables and Riverstone, who we know to be two of the most sophisticated energy investors in the nation. If exploration is successful and a viable commercial resource is found during the exploratory drilling, Davenport will prepare and submit specific plans to develop a state-of-the-art geothermal energy facility that will provide clean, renewable energy."
>
> Thomas King, Partner at US Renewables Group, said: "US Renewables Group is excited to partner with the Davenport team and to bring its resources to the development of what many believe to be the largest untapped geothermal resource in the US."

> Stephen J. Schaefer, Managing Director at Riverstone Holdings, said, "We are delighted to partner with the very experienced Davenport team to develop this new renewable resource. This project's potential output will help to satisfy the growing demand for affordable and reliable renewable power for the western region of the United States."

5.8 Demonstration Project in Partnership with AltaRock Energy

In October of 2009, Davenport Power received notification from DOE that two grant applications for exploration and demonstration projects were selected for negotiation and finalization under The American Recovery and Reinvestment Act of 2009 (ARRA). "The grant award for these two projects supports Davenport Power's belief that Newberry and our geothermal project hold significant promise for testing innovative technologies and for producing electric power. Davenport Power looks forward to the opportunity these efforts may hold for expanding America's and Oregon's renewable energy portfolio," stated Doug Perry, President of Davenport; "these represent additional steps in our rigorous and lengthy exploration phase at Newberry to utilize its tremendous geothermal energy potential to produce base-load power that is renewable and clean" (ThinkGeoEnergy 2010).

Davenport faced the challenge of not having access to water on the wells they drilled. The demonstration project to be implemented in partnership with AltaRock Energy would potentially overcome this challenge. It worked in the following manner (ThinkGeoEnergy 2010):

> A deep well, drilled last year to about 10,000 feet, will be used to create an engineered geothermal system (aka "Enhanced Geothermal System" or "EGS"). After exhaustive study and planning a network of small cracks/fractures is created surrounding the bottom of the well by injecting a controlled stream of cold water down the well into deep hot brittle rock. The resulting network of small cracks acts like a reservoir that can be filled with water. This water is heated as it passes over the surrounding hot rock to temperatures high enough for geothermal energy production upon its return to the surface through nearby production wells. Once this heated fluid is used for power production, it is re-injected into the engineered reservoir of small cracks to be reheated, thus creating a relatively continuous heating/power production cycle.

The U.S. Department of Energy described EGS as (AltaRock Energy/Davenport Newberry 2010):

> …extracting heat from the earth by creating a subsurface fracture system and circulating water through these fractures using deep well bores. Creating an EGS reservoir requires improving the natural permeability of rock. Rock is permeable due to the presence of minute fractures or pore spaces. Water pumped into deep injection wells is heated by contact with the rock and returns to the surface through production wells, similar to naturally occurring geothermal systems.

The Newberry Demonstration Project aimed at providing a model to be used at other similar sites; it was also important because it showed that if only 10% of the geothermal energy potential present in the Cascades Range could be harnessed, it could supply the entire Pacific Northwest with as much energy as five nuclear plants (Petty 2010). One of the previously drilled wells did not require any modification for the implementation of the EGS project because it exhibited low permeability and had a temperature of over 600 °F at total depth. This well represented an excellent opportunity for demonstration of EGS stimulation technology and it created the opportunity to expand the size of the geothermal resources available for recovery, both at Newberry and at other sites in the Cascades (Petty 2010).

Although one of the wells would be used for the Enhanced Geothermal Systems project, other structures and activities would have to be installed, according to the NOI. These included (Bureau of Land Management, 2011b) the following:

- Drilling three new MSA boreholes
- Installation and calibration of the final microseismic array
- Stimulation and testing of the injection well
- Drilling, stimulation, and testing of two production wells
- Long-term circulation test (30–60 days to collect data that would help create a conceptual model of how hypothetical EGS power plan might function)

5.9 Environmental Assessment

The BLM prepared an environmental assessment in order to disclose and analyze the environmental effects of developing and testing EGS technologies. This EA started when Davenport-AltaRock submitted an NOI and "Plan of Exploration, Operations Plan, and Drilling Program for the Newberry Volcano EGS Demonstration Project" to the Prineville BLM office in May of 2010. The BLM became involved because the NGP is located on federal geothermal leases administered by the BLM; also, the BLM is the lead agency because the majority of NGP activities would occur on leases issued and administered by the BLM (Bureau of Land Management 2011b).

Because the NGP was located on National Forest System lands (Deschutes National Forest) and because some monitoring stations were within lands where surface disturbance is under its authority, the Forest Service also needed to complete an EA. Therefore, this agency cooperated with the BLM in the assessment. In addition, DOE cooperated in this effort because the Energy Policy Act of 2005 gave the U.S. Secretary of Energy the authority to conduct a program of research, development, demonstration, and commercial application for geothermal energy, and DOE opted for funding a portion of the EA research project (Bureau of Land Management 2011b, pp. 12–16). The federal agencies involved are shown in the Box 5.1.

Box 5.1 Organizations Involved in the Environmental Analysis

BLM

"The proposed action would assist the BLM in meeting the management objectives in the Energy Policy Act of 2005 (Title II, Section 211), which establish a goal for the Secretary of the Interior to approve 10,000 MWs of electricity from non-hydropower renewable energy projects located on public lands. The proposed action also would further the purpose of Secretarial Order 3285 (March 11, 2009) that established the development of environmentally responsible renewable energy as a priority for the Department of the Interior."

Forest Service

"The FS will help the BLM in granting, granting with modifications or denying the implementation of the EGS. In addition: "A secondary purpose is for the Forest Service to carry out federal energy policy, orders, and objectives, including those from: (1) The National Energy Policy (May 2001) which includes the need "to expedite projects that will increase the production, transmission, or conservation of energy" (Section 1, Policy and Executive Order 13212), and (2) The Energy Policy Act of 2005 (Public Law 109–58) for promoting the leasing and development of geothermal resources where appropriate on public lands."

"As the manager of the national forest lands on which the EGS Project is proposed, Forest Service has a need to cooperate with BLM as it evaluates projects on geothermal leases that were issued with Forest Service consent. Forest Service also has a need to ensure that the proposed EGS Project meets the goals, objectives, standards and guidelines of the 1990 LRMP for the Deschutes National Forest, and the Newberry National Volcanic Monument Plan. This project has MSA stations being proposed where the Forest Service has the decision authority under NEPA and as a result the Forest Service has a need to make a decision to issue or not issue a permit for these stations."

DOE

"The National Environmental Policy Act (42 U.S.C. 4341 et seq.; NEPA), the Council on Environmental Quality's NEPA regulations (40 Code of Federal Regulations [CFR] Parts 1500–1508), and the DOE's NEPA implementing procedures (10 CFR Part 1021) require that DOE consider the potential environmental impacts of a proposed action before making a decision. This requirement applies to decisions about whether to provide different types of financial assistance to private entities."

"As part of the American Recovery and Reinvestment Act of 2009, DOE issued a funding opportunity announcement DE-PE36-09GO99019 Enhanced Geothermal Systems Demonstrations through which DOE sought to fund projects in a variety of geologic formations that could quantitatively

(continued)

Box 5.1 (continued)
demonstrate and validate stimulation techniques that successfully sustain sufficient fluid flow and heat extraction rates for 5–7 years that produce up to 50 MWe per year per project site/geothermal reservoir. The proposed EGS Project has the potential to advance EGS technology by developing and testing an EGS reservoir. DOE is proposing to authorize the expenditure of federal funding through the GTP to AltaRock for the proposed EGS Project. DOE has already authorized the use of a small percentage of the Federal funding for preliminary activities and associated analyses."

The EA was to be an evaluation of the Development Project only, not including the implementation of a plant. Some of the actors involved in the environmental assessment included, but were not limited to, the Confederated Tribes of the Warm Springs Reservation, Klamath Tribes, Burns Paiute Tribes, USDA Forest Service Regional and National Offices, BLM National Office, and the URS Corporation regarding seismic risks (see also the box, "Organizations Involved in the Environmental Analysis"). The assessment, released in 2011, was also based on previous environmental analyses performed according to NEPA standards. A first detailed analysis near the EGS area was conducted in 1994 by Deschutes FS, BLM, and Bonneville Power Administration (BPA) with inconclusive results; the second was performed in 2007 regarding the Davenport drilling area (Bureau of Land Management 2011b: 22).

5.10 Public Participation in the Definition of Issues

Between July and November 2010, several public meetings were held to provide information about the proposed project at Newberry, with the participation of BLM, Forest Service, DOE, AltaRock, and Davenport. The meetings were meant to inform and engage the broadest possible central Oregon audience. In July, a meeting in La Pine gathered 25 individuals; in August, a meeting in Sunriver gathered more than a dozen people. Another 26 people attended a public meeting held in Bend in September. The BLM, the Forest Service, and NGP proponents led a field tour during the public scoping period on November 10, 2010, with 25 members of the public participating (Bureau of Land Management 2011b).

The BLM received 11 comments following publication of the draft EA; these comments identified several primary concerns regarding toxicity, seismicity, and groundwater connectivity, as well as several other issues. The letters came from the Klamath Tribe, Region 10 of the Environmental Protection Agency, the Northwest Environmental Defense Center, Oregon Wild, Central Oregon Land Watch, Blue Mountain Biodiversity Project, Leaning Pine Ranch, and four individuals.

Concerns and topics raised by the experts from cooperating agencies were reviewed, used to develop "key issues," and to help guide the Final Environmental Assessment. Some issues covered in public comments and letters were not analyzed in detail because they were not considered key issues. These issues were: disturbance of forest vegetation and timber stands; effects to socioeconomics; effects to infrastructure; utilities, energy and materials; effects to transportation; environmental justice; management of health safety and hazardous materials; effects to cultural resources; disturbance to threatened, endangered, and sensitive plants; effects to noise levels; effects to air quality; effects to land use; effects to wilderness areas; inventoried roadless areas; wild and scenic rivers; impact on winter recreation; irreversible and irretrievable commitment of resources; and intentional destructive acts.

5.11 Issues Addressed in the Environmental Assessment

5.11.1 EGS Safety

By the time the Newberry Crater demonstration started, AltaRock had not yet tested technology; although they were working in northern California, the project was halted before EGS was implemented due to drilling difficulties. Despite that, AltaRock and Davenport indicated they had assembled a very qualified team of scientists and engineers and that they were following a "prudent approach" in order to guarantee safety (AltaRock Energy/Davenport Newberry 2010). When questioned about the safety of EGS, AltaRock and Davenport provided the following answer (Bureau of Land Management 2011b, p. 23–28):

> Several hot dry rock and EGS projects have been conducted over the past 30 years, advancing the technology in degrees. There has, however, been one aberrant project over the last three decades. An attempt to create an EGS reservoir in the bedrock beneath Basel, Switzerland, caused an earthquake that registered 3.4 on the Richter Scale. This was a very small disturbance. Nobody was injured and there was only minor structural damage. Much of the controversy and damage was due to the location in the midst of a densely populated area with older (and, in some instances, centuries' old) large buildings. Compounding the reaction was the fact that people were not prepared for the possibility of such events. We're not being defensive, and we're not trying to be "Monday morning quarterbacks," but the project in Basel drilled and injected into an active earthquake fault in a densely populated area. Like virtually all the other proponents and practitioners of EGS, AltaRock does not target known faults and, in fact, avoids them; it also takes great pains to notify and educate the community about all its activities up front. [...]Unlike the Basel project, the Newberry project is a safe distance from the nearest residents, and the area is not densely populated. In fact, the project is about 6.5 miles from the nearest private residence.

5.11.2 Water Consumption and Threats to Water Quality

The EA identified two risks regarding water use: (1) "Withdrawal of groundwater from water wells for the development and testing of a below ground EGS reservoir has the potential to reduce the quantity of water available for other uses within the Deschutes drainage basin" (Bureau of Land Management 2011b: 29); and (2) "The development and testing of a belowground EGS reservoir has the potential to negatively impact groundwater quality within the aquifer" (Bureau of Land Management 2011b, p. 30).

Groundwater was going to be used for the demonstration project; the water would be pumped from shallow wells located at the project site to create the EGS reservoir, drill two production wells, and then extract the geothermal energy. According to NGP, adequate groundwater supplies are available at the site to support the demonstration (AltaRock Energy/Davenport Newberry 2010). Two water wells already existed on site and two more were likely to be added. According to AltaRock President Susan Petty, "the wells are designed and completed with at least three strings of steel casing cemented in place to block any communication between the inside of the well and groundwater resources" (Petty 2010: 8). While wells were developed, the location of fractures was constantly monitored; stimulation pumping was stopped when the fractures extended far enough from the well to get the right heat exchange surface, a radius of about 1640 feet. At this distance, the fractures would not affect any usable groundwater because the shallowest depth at which the fractures can start is at about 6500 feet; also, studies show that there is no fresh groundwater deeper than about 1000 feet (Petty 2010). Water used in the demonstration project for the stimulation, drilling, and testing was also provided through the purchase of temporary water credits from the Deschutes River Conservancy, pursuant to Oregon Water Resources Department rules and the Deschutes Groundwater Mitigation Bank procedures. Davenport's temporary credit purchases helped support the Mitigation Bank and Middle Deschutes River restoration (ThinkGeoEnergy 2010).

Multiple mitigation measures were implemented to minimize the possible impact of project activities. Thus, all drilling wastes would be contained in the existing lined drilling sump and disposed of promptly and properly, and any potential runoff from the drill pad would also be directed to the sump; drilling fluids were formulated from non-toxic components; and all non-geothermal groundwater was cased off and cemented (AltaRock Energy/Davenport Newberry 2010, p. 3).

Newberry Geothermal Project representatives stated the following in October of 2009:

> [I]n the longer term the amount of water – or an alternative – that ultimately will be needed for operating EGS, or even traditional hydrothermal geothermal system plants at Newberry that transports heat to the surface will depend on a number of factors that are not known at this time. These factors include plant size (number of MW's), plant type and efficiency, and whether or not there will be any fluid losses, etc. from any EGS reservoirs. It is possible that over time there may be a mix of options, such as liquefied CO2 and other technologies that

> are under development and testing, to transport heat from EGS reservoirs deep in the earth to the surface so that less water will be needed. (ThinkGeoEnergy 2010)

Another area of concern regarding water conservation is the protection of Paulina Lake and East Lake, which are located in the Newberry Monument. AltaRock President Petty (2010) stated that the lakes sit at a higher elevation (6350 feet) than the project site (5815 feet); therefore, fluids from the stimulation and drilling operations cannot flow up into and affect the lake.

5.11.3 Earthquakes

Another risk identified in the EA was that "the development of a below ground EGS reservoir by hydroshearing has the potential to produce induced seismicity and increased seismic risk that could affect historic structures, resorts, and other recreation sites within the NNVM, could increase avalanche risk, could increase risk to above and below ground geologic features, and could result in property damage in nearby population centers" (Bureau of Land Management 2011b, p. 30).

In 2010, Newberry's representatives announced that stimulating fractures using EGS technology does create small seismic events, or "micro seismicity"; however, these events were said to likely be below the human detectable level. It was stated also that it was undesirable from an engineering standpoint to create larger fractures that might result in seismic events that can be felt by nearby residents. As a safety measure, regulators would monitor the activity and the pumping could be suspended at any moment. In addition, seismic data recorded at the site would be streamed real-time to Lawrence Berkeley National Laboratory, where it would be processed and displayed on a public web site (AltaRock Energy/Davenport Newberry 2010).

According to Petty (2010), several factors affect the possibility of generating seismic events that could be felt by nearby communities, such as:

- **Close proximity of the stimulated volume to large scale, tectonically-active faults:** None have been mapped so far at Newberry, but they proposed to use sensitive LiDAR mapping methods to be sure that there are none of these active faults in the area.
- **A past history of larger seismic events:** There have only been three seismic events recorded by the USGS in this area over the last 10 years, all of them too small to be felt by local residents.
- **Pumping of large volumes of fluid at pressures that exceed the failure pressure for the rock:** The stimulation methods to be used at NGP would rely on lower pressures, below the failure pressure for the rock, so that bigger fractures were not created.
- **A big imbalance between production and injection volumes and pressures:** The project is designed to produce everything that was injected on a "closed loop"; this not only reduces the risk of big pressure differences, but also helps conserve water.

AltaRock submitted an Induced Seismicity Mitigation Plan (ISMP) prepared for the Newberry Enhanced Geothermal System (EGS) Demonstration, which was approved on August 10, 2011 by DOE's Geothermal Technology Program. The ISMP followed a protocol developed by the International Energy Agency (IEA) for addressing induced seismicity during geothermal projects that was adopted by DOE for EGS demonstration projects (AltaRock Energy 2011b).

For support for the ISMP, AltaRock used an "evaluation of induced seismicity/seismic hazards and risk" prepared by an independent agency in 2010 (AltaRock Energy 2010). This evaluation concluded that potential EGS induced seismicity poses no seismic risk to the residents in the neighboring communities. More specifically it states the following:

> Although there is no additional seismic risk due to EGS induced earthquakes, if events of M 3.0 and higher were to occur, and we judge the likelihood of their occurrence to be small, they will probably be felt in La Pine and Sunriver, but not at damaging levels of ground motions (> 0.10 g). Individual residents within 10 km of the Project site will feel the larger events. The strength of shaking will depend on the size of the event, and distance to and site conditions at each location. The effects of induced seismicity will be more of a nuisance than a hazard to the vast majority of local residents because of the small size of the events and distances to centers of population. (AltaRock Energy 2010: 1)

On their ISMP, AltaRock stated that they believe "the model used by URS Corporation to evaluate risk, which is based on data from The Geysers geothermal field, overestimates the shaking that might occur at Newberry, and thus represents a cautious approach" (AltaRock Energy 2011b, p. 2).

The purpose of the ISMP was to demonstrate AltaRock's preparedness for addressing and mitigating potential seismic activity of concern to people at the demonstration site. "The Department of Energy's approval of our Induced Seismicity Mitigation Plan highlights the great level of care and attention we have brought to this issue since the project's inception," said Will Osborn, Project Manager for AltaRock. "In our public outreach meetings we've received and responded to concerns about water usage and potential seismic activity; approval by the Department of Energy affirms that this project is well-planned and well-executed, and we're excited to keep the process moving along here in Oregon. We encourage the public to review the approved plan available on our web site, and continue to provide comments and questions through our new toll-free information line" (AltaRock Energy 2011a, p. 1).

In the ISMP, AltaRock stated the following in regard to mitigation measures: "For control and mitigation of induced seismicity, this document defines limits (or 'triggers') that, if activated, will initiate mitigation actions up to and including stopping injection and immediately flowing the well to reduce reservoir pressure. The triggers will be monitored during hydroshearing and EGS reservoir creation, and throughout the remainder of the Demonstration." These triggers were based on real time measurement of seismic activity on the PNSN regional network, the AltaRock MSA, and the Paulina Lake SMS.

5.11.4 Impacts to the Newberry National Volcanic Monument

The Newberry National Volcanic Monument is considered a national treasure. The central caldera region of the volcano was designated a national monument through the efforts of a committee formed because of local grassroots efforts begun in 1989. This committee represented a wide range of interest groups, including the community, environmental groups, government, and the geothermal industry. In 1990, federal legislation designated the central caldera and a narrower band of land running north to Lava Butte as the Newberry National Volcanic Monument. The legislation also "included provisions for a geothermal lease swap, whereby existing Federal geothermal leases within the new National Monument were exchanged for leases *outside* of the boundary. This action excluded geothermal exploration and development within the newly created Monument, but provided that it may occur on leases outside the Monument." Therefore "the creation of the Monument was not intended to eliminate geothermal development in the region, but to ensure it is done in the most appropriate and safe location" (AltaRock Energy/Davenport Newberry 2010). Since the leases were transferred to outside the monument, there have been four high-temperature, low-permeability wells drilled at the Newberry site, which confirmed that there is a large heat resource in the northwest portion of the leased area (Petty 2010).

5.11.5 Visual Impact

According to the EA, the visual impact risk was as follows:

> [The] removal of vegetation on the microseismic monitoring sites has the potential to cause up to 3 areas of approximately 9,375 square feet (0.2 acre) each or a total of 28,125 feet (2/3 acres) to not meet the Forest Plan standards for visual quality as seen from selected viewpoints. The Deschutes LRMP Standards and Guidelines that supports this issue statement is M8-19. The venting of steam during the short and long term circulation tests may also create a steam plume that could potentially be visible at times from certain selected viewpoints. The drill rig and circulation testing facilities may be visible at times from some key viewer locations during the anticipated 2-year duration of the Project. (Bureau of Land Management 2011b, p. 30)

However, the EGS Project was located in an area that had already been involved in geothermal exploration, including well drilling, since the 1980s. The area has also been used for logging in the recent past. These activities have already shaped the appearance of the landscape (AltaRock Energy/Davenport Newberry 2010). In addition, it is important to note that all demonstration project installations were to be temporary. The temporary drilling rig mast would be approximately 140 feet in height and might have been visible from Paulina Peak. A steam plume resulting from drilling and flow testing of the wells (that would occur for no more than a total of approximately 80 days over a period of three years) would be visible from nearby viewing points, particularly during cooler seasons. Placement of equipment, such as that associated with the microseismic array (MSA), would be small in size and carefully sited. "Distance, topography, heights of surrounding trees, and existing landscape variation will also contribute to screening the project and making the operations not readily seen from roads or viewpoints, either inside or outside of the Monument" (AltaRock Energy/Davenport Newberry 2010).

Finally, most operations would occur on existing well pads or below ground, resulting in little new surface disturbance. Where surface disturbance might occur, sites with the least amount of vegetation would be utilized, such as skid trails and landings from past logging operations (AltaRock Energy/Davenport Newberry 2010). A "Scenic Resources Inventory and Assessment" conducted by PLS Environmental, in February of 2011, analyzed the impact of the proposed action and of an Alternative B; it found that "impacts to scenic resources resulting from either action alternative would not be significant." They also proposed mitigation measures based on three criteria: (1) strategic location, (2) minimization of disturbance, and (3) repetition of the basic landscape elements (form, line, color, texture) (PLS Environmental 2011, p. 19).

5.11.6 Wildlife – Flora and Fauna

A Biological Evaluation regarding Threatened, Endangered and Sensitive Plants (TES plants) analyzed the effects of the Project on TES plant species that are documented or suspected to occur on the Deschutes National Forest. The results of the report, published in August of 2010, found that the proposed action would have no impact on these plant species because the project area either did not contain them or had a low probability for being their habitat (Cooper 2010).

However, there is potentially suitable habitat near the proposed project area for Cooper's hawk, northern goshawk, sharp-shinned hawk, red-tailed hawk, northern flicker, hairy woodpecker, and flammulated owl. The drilling activities may disrupt or remove nesting, foraging, and fledging habitat at the well pads and proposed MSA borehole locations; however, "any disturbance to a known nesting pair within ¼ mile of a work location will be mitigated by halting operations during the breeding season (e.g., March 1- Aug 31 for goshawk)" (AltaRock Energy/Davenport Newberry 2010).

According to the EA, the wildlife issue identified was that "preparing and clearing the vegetation for the three borehole MSA stations has the potential to remove habitat on these sites for some species. Drilling activities, testing, and stimulation activities, and an increase in human disturbance also have the potential to disturb nesting sites up to ¼ mile during the breeding season or temporarily displace some wildlife species" (Bureau of Land Management 2011b, p. 29).

5.12 Decision Record and Finding of No Significant Impact

The EA provided the analysis used by BLM to determine whether it could issue a Finding of No Significant Impact (FONSI) or whether it was necessary to prepare an Environmental Impact Statement (EIS) according to NEPA standards. If the BLM determined that the project project did not present significant effects, based on the EA, a Decision Record (DR) would be issued approving a selected alternative, whether it was the Proposed Action or another alternative. The Forest Service would make a separate NEPA decision and issue a FONSI, or declare the need to prepare an EIS, on those nine MSA stations that were not on geothermal leases administered by BLM. The DOE would make the same decision, in this case in regards to allowing the expenditure of federal grant money (Bureau of Land Management 2011a).

On April 5, 2012, a DR and a FONSI were issued for the EGS project. In the FONSI, BLM Prineville District Manager Carol Benkosky stated:

> [B]ased upon a review of the EA and its supporting documents, I have determined that the project is not a major federal action that would significantly affect the quality of the human environment, individually or cumulatively with other actions in the general area. None of the environmental effects described for the proposed project would meet the definition of significance in terms of context or intensity, as defined in 40 CFR 1508.27 and described below, nor would they exceed those effects described in the Deschutes National Forest LRMP 1990, as amended. I have therefore determined that an environmental impact statement is not required for this project. (Benkosky 2012, p. 2)

In making this decision, District Manager Benkosky considered the proposal's consistency with NNVM Act (Public Law 101–522), the proposal of installing a micro-seismic monitoring station and one strong motion sensor within the NNVM, and the detailed ISMP that was developed to assess and mitigate any potential effects that could result from seismicity induced by the project. In the FONSI, Benkosky suggested that "Given that the physical effects of the project are limited to the local geographic area, the project's consistency with the NNVM Act, and the mitigation measures contained in the ISMP, I find that the EA has not identified any direct, indirect, or cumulative effects of sufficient size or duration to be significant at the local, regional, or national level. If successful, the project could have important implications for future development of geothermal resources in a broader context" (Benkosky 2012, p. 3).

The DR stated that "after a review of the comments received, and based on the Environmental Assessment (DOIBLM-OR-P000-2011-0003-EA) and the Finding of No Significant Impact, it is my decision to allow Davenport Newberry Holdings LLC ("Davenport") and AltaRock Energy, Inc. (collectively the "Proponents") to develop and test an Enhanced Geothermal System (EGS) ("Project") as described in Alternative A of the EA" (Bureau of Land Management 2012, p. 1). The duration of implementation and testing of EGS was foreseen to last two years. The rationale for approval of the Newberry Geothermal Project is as follows (Bureau of Land Management 2012):

> This decision is consistent with the purpose for which lands were leased by the United States to Davenport Newberry Holdings LLC, and which conveyed to Davenport Newberry the "exclusive right and privilege to drill for, extract, produce, remove, utilize, sell, and dispose of geothermal steam and associated geothermal resources." To maintain this right, Davenport Newberry must "diligently explore the leased lands for geothermal resources…" applicable to each of these leases.
>
> The Geothermal Steam Act of 1970, as amended, 30 U.S.C. 1001-1028, provides the authority for the BLM to allow for the exploration, development, and utilization of geothermal resources on BLM-managed public lands, as well as geothermal resources on lands managed by other surface management agencies, such as the United States Forest Service.
>
> As a cooperating agency, the Forest Service has fully and actively participated in the environmental analysis process for the Project, has prepared resource analyses for the EA, and has reviewed all studies and documents associated with the NEPA process.
>
> The decision is consistent with surface use stipulations that were made part of the Leases OR40497, OR45505, OR65371, OR12399, OR12437, OR65470, OR45506, OR47300, and OR53085 held by Davenport Newberry Holdings, LLC.
>
> The exploration of the geothermal resource is consistent with initiatives of the National Energy Policy and supports the National Renewable Energy Initiative by providing more information about the energy production from geothermal resources.
>
> No impacts were identified in the EA that will not be adequately mitigated by design features built into the Project including the required Induced Seismicity Mitigation Plan (both summarized earlier in this Decision Record).

5.13 Sundry Notice from the BLM – 2012

Newberry EGS announced in September 2012 that the Newberry EGS Demonstration had obtained a "sundry notice" approval from the BLM. The notice allowed the stimulation process to create geothermal reservoirs to begin, which was part of Phase II of the project. The stimulation was scheduled to begin in October 2012 (KTVZ 2012). AltaRock President Susan Petty said the BLM approval would allow her company to see how much potential for geothermal energy existed at the site and whether it would be worthwhile to build a power plant there; if the demonstration promised at least 10 megawatts of power, it would justify further action toward developing geothermal energy project. "If it were economic, then we would start permitting to build a power plant," said Petty. "But only if it were economic. Only if it were successful. Success means we can access enough of the hot rock down there with small fractures to be able to flow at economic rates and get enough water to production wells to be economic. That's the part that we just don't know at this point" (Profita 2012c).

5.14 Controversy Regarding the Project

Controversy over different aspects of the geothermal project began to emerge after the approval from the BLM in 2012. The concerns centered on possible chemical contamination from the project, potential earthquakes caused by project activities, the impacts of "fracking" near a volcano, and the economic viability of the project.

5.14.1 Chemical Controversy

Examples of concerns expressed about chemicals that might be used in the project, along with responses from project partners, are shown below.

> A family doctor in Bend who helped create Central Oregon's Newberry Volcano National Monument has raised concerns about the chemicals that would be injected underground as part of a proposed geothermal energy project near the volcano [...] 'I'm concerned that there hasn't been full disclosure. There hasn't been transparency here. And I think people in Central Oregon having to deal with potential effects of the drilling need to be fully informed,' says Garrett. [...] AltaRock Energy President Susan Petty said the company has deliberately chosen chemical agents that are not toxic. She said the chemicals would be used in very minute quantities far below the groundwater table and precautions would be taken to avoid any leakage (Profita 2012a).

Vince Patton, an OPB reporter, asked Will Osborn, the project manager for AltaRock, about whether the process to be used was "similar to the controversial hydraulic fracturing action [...] Osborn says there's a big difference: 'In hydro-fracking done in the oil and gas industry, they actually shatter the rock. They break the rock and create new cracks, new fractures in the rock. We just inject cold, clean groundwater,' Osborn says. 'In hydrofracking they use a mixture of chemicals.' [...] But then, as Bend family doctor Stuart Garrett recently discovered, AltaRock will, in fact, be using its own mixture of chemicals in the injection process. So, what gives? Is this fracking or not?" (Profita 2012b). A piece of Mr. Garrett's response to the list of chemicals that would be used by AltaRock was also published:

> Davenport/Altarock has not publicly announced that it will be pumping industrial chemicals (tracers, diverters, mud additives, friction reducers, etc) including naphthalene, safranin, rhodamine, lithium, cesium, rubidium, fluorescein, tobermorite, polymerized plastics, secret Altavert fluids, and other compounds into the wells. This has not been reported in public meetings, in written materials, or in news articles until the EA came out. More transparency and candor is called for. [...] While it has been asserted that the proprietary Altavert materials are harmless, the public deserves to know what they are and to make their own judgment. Secrecy does not belong in this EA (Profita 2012b).

In response to these comments, Linda Christian, the Newberry project manager for BLM, backed up arguments provided by AltaRock representatives:

> She said her agency will know the individual chemicals being used in the proprietary products, which are 'like biodegradable plastic bags' that degrade when they get near the heat deep underground. [...] 'It's an understandable response,' she said. 'But hydroshearing isn't much like hydrofracking. Fracking requires chemicals or sand to hold the fractures open, and that is not at all what's going to happen with hydroshearing. These cracks already exist. What they may not be is continuous. By adding pressure, mostly with millions of gallons of water going down there, the extra pressure makes the rocks move, so it becomes a continuous conduit of water over hot rock like a heat exchanger (Profita 2012b).

5.14.2 Earthquake Controversy

There were also concerns over the threat of earthquakes associated with the project. In August of the 2012, when AltaRock was involved in the process of acquiring the Sundry Notice, *Forbes* had announced: "The pilot project is underway at Newberry Volcano despite concerns that have dogged other "enhanced geothermal" projects, such as small earthquakes, and techniques that are similar to the controversial practice of fracking. **Oh, and there's the landscape's habit of exploding**" (Ferris 2012).

In 2013, a review about AltaRock President Petty was published and, recalling her actions in California, stated: "With a team of scientists and engineers (and both private and public funding in hand), Petty attempted to build an enhanced geothermal energy project at The Geysers, a geothermal hotspot just north of San Francisco. Everything seemed to be going right. The Obama administration even positioned AltaRock as a paragon of American energy independence – a startup U.S. company working toward a brighter, greener future" (Lecher 2013). Petty's work was compared to the one developed by the Basel Geothermal Project in Switzerland:

> AltaRock's timing was decidedly inauspicious. In December 2006, earthquakes started near an enhanced geothermal project in Basel, Switzerland. Markus O. Häring, a former geologist and oilman who started the geothermal power company Geopower Basel, tried to shut down the operation, but it was too late. Earthquakes rocked the area for a year, doing almost $9 million worth of damage. Häring faced criminal charges for the damage, potentially facing up to five years in prison as prosecutors argued he downplayed the risks (he has later acquitted). He was quoted as saying the people in the area knew the risks, and that the quakes were "a learning process for everyone involved [...] Before long, AltaRock was being portrayed as the bad guy—a super-villain out to damage surrounding areas. The Energy Department and the Bureau of Land Management ordered a new review of the safety of The Geysers project. Eventually the project was approved anyway, but, plagued by technical problems, Obama's paragon of American energy independence decided to head north, to Oregon, instead. (Lecher 2013)

The New York Times ran an article about AltaRock's project in northern California, while the company was already starting the permit processes in Oregon, saying: "in a report on seismic impact that AltaRock was required to file, the company failed to mention that the Basel program was shut down because of the earthquake it caused. AltaRock claimed it was uncertain that the project had caused the quake, even though Swiss government seismologists and officials on the Basel project agreed that it did. Nor did AltaRock mention the thousands of smaller earthquakes induced by the Basel project that continued for months after it shut down" (Glanz 2009).

"Could the process be 100-percent guaranteed safe? 'There's never guarantees in any kind of geothermal situation,' Petty says, while insisting that the safety risks are so low, they amount to more of a rounding error than a serious threat. As for the Basel incident, she says that's something else entirely: 'What they were doing was more like oil and gas.' The water pressure of the Basel project, she says, was higher than AltaRock's" (Lecher 2013).

A positive review was also published in 2012: "Still, the DOE reports that Geysers in northern California, a geothermal project that produces more than 700 megawatts of electricity, has caused countless microseismic events and one earthquake of magnitude 4.6. These events have all been measured at less than the size of a magnitude 5 quake that would be deemed dangerous. Given that the project at Newberry is taking place on a much smaller scale, there's probably not too much to worry about. Perhaps oil and other fossil fuel companies are the only ones who should be concerned about AltaRock's vision for the future" (Matteis 2012).

5.14.3 Fracking Near a Volcano Controversy

Another issue that caused much controversy was the potential impact of drilling and fracking-like activities near a volcano. Some examples of the concerns expressed about "fracking" are as follows:

> No matter what it may sound like, Susan Petty insists she's not trying to frack a volcano. Petty, the founder and president of the Seattle startup AltaRock Energy Inc., says she is just trying to do for geothermal energy what hydraulic fracturing has done for oil and gas. About two months from now, the company will pump millions of gallons of water into the ground near the base of an Oregon volcano using Petty's version of a technique that Halliburton Co. first used on a geothermal well in the 1970s. Its given name? Hydraulic fracturing. Fracking being as controversial as it is, Petty prefers to call it 'hydro-shearing.' AltaRock says there are major differences between the two techniques, and Petty has convinced three government agencies that there is no real risk of water contamination, earthquakes or volcanic activity from its project. Now the company must prove to the business world that Petty's patented method can deliver cheap energy [...] So she came up with a technique of her own. It was so tantalizing that AltaRock managed to raise millions of dollars from the Internet titan Google Inc., the closely watched venture capital firm Kleiner Perkins Caufield & Byers and the Department of Energy, all around the time that the company celebrated its first birthday in 2008. (Nelson 2012)
>
> Fracking is a method typically used to extract fossil fuels, and one that has caused controversy in recent years because of its potential to trigger (albeit mild) earthquakes. According to a recent survey by the Royal Society and the Royal Academy of Engineering, however, earthquakes are unlikely if companies employ "best practice" and enforce strict regulatory procedures. With any risk factor at all, however, it seems odd that a volcano would be a first choice to set up shop. Especially when that volcano hasn't erupted for 1,400 years and the U.S. geological survey says it "is certain to erupt again. (Matteis 2012)

Some of the concerns focused on the project's location near a volcano:

> Some of the claims, like the idea that the volcano will erupt because of the project, don't have much basis in science (although the volcano went through a period of dozens of eruptions thousands of years ago, and 'is certain to erupt again,' according to the U.S. Geological Survey). As AltaRock is quick to point out, it's not pumping directly into the volcano, but nearby it. But the earthquake link was enough for the government to order a study and risk analysis of the project. URS Corporation, an independent body, released its verdict: any risk was low, and the 'effects of induced seismicity will be more of a nuisance than a hazard to the vast majority of local residents because of the small size of the events and distances to centers of population. (Lecher 2013).

> Injecting water at a very high pressure into the cracks of a dormant volcano in order to exploit an untapped source of clean energy? Does that sound like a good idea? [...] The idea of the $44 million project is to break up the rock surrounding the volcano by cracking it apart with high pressure water and the many other substances that go into fracking to make the geothermal heat contained in those rocks easier to access. But wait, hasn't fracking been linked with earthquakes in Oklahoma and Ohio recently? And now the BLM has approved fracking a volcano? I'm no scientist, but this sounds like a bad idea. [...] Geothermal energy sounds like a great idea to help solve our energy challenges and fight global warming, but maybe not fracking a volcano. (Molland 2012)

AltaRock responded to concerns about the location near the volcano, as well as to comparisons of EGS and fracking as follows:

1. **AltaRock Energy is not drilling into a volcano.** The well on Newberry was drilled in 2008 to a depth of over 10,000 feet. about 5 miles from the most recent volcanic flows. We are pumping cold water into this non-producing geothermal well in an attempt to prove new technology for Enhanced Geothermal Systems developed by AltaRock and many others, and eventually generate renewable energy in central Oregon. We did drill five seismic monitoring boreholes to depths of 700–900 feet, but the temperatures at the bottom of those wells is cold (55F) and their purpose is monitoring not geothermal production.
2. **Drilling into volcanoes is not dangerous or uncommon.** Worldwide geothermal energy production often occurs on volcanoes. The Philippines, Iceland, and El Salvador all produce over 20% of their electricity from the geothermal heat of volcanoes. Geothermal drilling or production has never been responsible for a volcanic eruption.
3. **We aren't "fracking."** Hydraulic fracturing is a process used in the oil and gas industry that uses surface water pressures over 5000 psi to fracture or crack open the rock to extract oil and gas. We are using a different process called hydroshearing to initiate small slips in existing cracks in the rock. We expect surface pressures around 1600 psi will initiate hydroshearing, or slip on existing fractures — far less than needed to fracture the rock. Further, hydroshearing does not require injection of the chemical additives used in fracking. Lastly, we will not be producing fossil fuels, but hot water and steam heated by the hot rock under Newberry Volcano" (Newberry Geothermal Energy 2012).

5.14.4 Economic Viability Controversy

Finally, there was controversy concerning the economic viability of the project. Some representative examples of concerns and responses are shown below.

> AltaRock claims that by creating a large network of cracks at multiple depths, it can lower the price of an enhanced geothermal plant, which is estimated to costs up to three times as much as a coal plant of similar size [...] "What we think we can do at Newberry is at least double the production from each well. If we could double the production of the well we could cut the cost of production from 30 percent to 35 percent. If we are really successful,

we could cut it by 40 percent to 45 percent and that would make this a really economically successful proposition." (Ferris 2012)

"The economic implications of the Newberry EGS Demonstration are quite significant," said Susan Petty, president and founder of AltaRock Energy. "Creating multiple geothermal reservoirs from a single injection well multiplies the amount of energy that can be extracted from each production well [...] Not only do multiple reservoirs extend the life of the well and increase the energy recovery from each well, they significantly improve the economics of EGS power generation. This could allow geothermal to translate into a comprehensive source of renewable energy for the future. (KTVZ 2012)

Just as fracking found an early use as a way to squeeze more oil and gas from wells that were starting to run dry, geothermal developers such as Calpine Corp., Ormat Technologies Inc. and U.S. Geothermal Inc. are looking for ways to make their existing wells more productive and to recoup the cost – usually $3 million to $10 million per well – of hitting nothing but dry rock. [...] Karl Gawell, executive director of the Geothermal Energy Association, said those sort of advances will help to lower the risk across the industry. Today's geothermal power plants all rely on large natural reservoirs of hot water, the biggest and hottest of which have mostly been tapped. If a developer drills down into a new hot spot and doesn't hit much water, it is as if an oil well came up dry. "It doesn't take too many $10 million dry holes for a company to get very discouraged about geothermal projects," Gawell said. (Nelson 2012)

In 2013, AltaRock achieved the development of the promised technology:

AltaRock Energy today announced that it has created multiple stimulated zones from a single wellbore at the Newberry Enhanced Geothermal System (EGS) Demonstration site. Creating multiple stimulated zones from a single well will dramatically increase the flow and energy output per well for the completed system, which will include soon-to-be-drilled production wells. The overall effect will be to lower the cost of geothermal energy production by as much as 50 percent. This is a major advancement for EGS and has the potential to move geothermal energy from a niche role to a major player in AltaRock's energy portfolio. (Business Wire 2013)

The purpose of the Newberry EGS project is to demonstrate AltaRock's new technology designed to lower the cost of EGS, and thus allow economic extraction of heat from the earth in locations where high temperatures can be reached by conventional drilling techniques. (Business Wire 2013)

Until now, a major challenge facing EGS projects has been the cost. Since geothermal wells can cost several million dollars to drill, increasing the generation capacity of each well is critically important for making it more competitive on the open market for electricity because it lowers the cost of the produced power by reducing the number of wells that need to be drilled. Systems with single stimulated zones simply don't yield enough power per well to make the electricity produced cost competitive. Having the ability to create multiple stimulated zones from a single well increases the amount of power produced by the well, and thus decreases the overall cost of the power produced. (Business Wire 2013)

5.15 Lessons Learned

While the development of geothermal energy at the Newbury Volcano is still in testing stages and in pursuit of additional resources for development, there remain public concerns regarding toxicity, seismicity, and groundwater connectivity, as well as several other issues. The Klamath Tribe, the Northwest Environmental Defense

Center, Oregon Wild, Central Oregon Land Watch, Blue Mountain Biodiversity Project, and other stakeholders are watching the project closely and are not so convinced that the environmental benefits of the renewable energy project outweigh the potential costs associated with the project. As with our other case studies, there is also concern about the credibility of the developer and the science underpinning the project. However, the project was not selected for development in August of 2016 and is currently on hold as the consortium of developers search for other funding sources for the project.

References

AltaRock/Davenport Power. (2007). Newberry geothermal project: What's next following approvals for exploratory test wells. *AltaRock Energy*. http://altarockenergy.com/projectupdates/NewberryQandA.pdf. Accessed 15 Aug 2016.

AltaRock Energy/Davenport Newberry. (2010). Newberry GES demonstration project: Frequently asked questions and answers. *AltaRock Energy*. http://altarockenergy.com/projects/newberry-egs-demonstration/. Accessed 15 Aug 2016.

AltaRock Energy, Inc. (2010). Final report – Evaluations of induced seismicity/seismic hazards and risk for the Newberry Volcano EGS demonstration. *AltaRock Energy*. http://www.altarock-energy.com/projectupdates/URS_SeismicReport.pdf. Accessed 15 Aug 2016.

AltaRock Energy, Inc. (2011a). Department of Energy approves Newberry EGS demonstration seismic mitigation plan. *AltaRock Energy*. http://altarockenergy.com/wp-content/uploads/2014/02/DOE-Approves-Newberry-EGS-Demo.pdf. Accessed 15 Aug 2016.

AltaRock Energy, Inc. (2011b). Newberry EGS demonstration induced seismicity mitigation plan. *AltaRock Energy*. http://www.altarockenergy.com/projectupdates/Newberry_ISMPlan.pdf. Accessed 15 Aug 2016.

Benkosky, C. (2012). Finding of no significant impact determination. Prineville District: Bureau of Land Management. http://www.blm.gov/or/districts/prineville/plans/newberry/files/Newberry_EGS_Demo_Project_FONSI.pdf. Accessed 10 July 2015.

Bureau of Land Management. (2011a). Newberry Volcano enhanced geothermal system (EGS) demonstration project environmental assessment. Washington, DC: US Department of Interior. Bureau of Land Management. http://www.blm.gov/or/districts/prineville/plans/newberryegs/. Accessed 15 Aug 2016.

Bureau of Land Management. (2011b). Wildlife report for the Newberry Volcano enhanced geothermal system (EGS) demonstration project. Washington, DC: US Department of Interior. Bureau of Land Management. http://www.blm.gov/or/districts/prineville/plans/files/Wildlife_Report.pdf. Accessed 15 Aug 2016.

Bureau of Land Management. (2012). Decision record. Washington, DC: US Department of Interior. Bureau of Land Management. http://www.blm.gov/or/districts/prineville/plans/newberry/files/Newberry_EGS_Demo_Project_Decision_Record.pdf. Accessed 15 Aug 2016.

Business Wire. (2008). Davenport power announces agreement with US renewables group and riverstone holdings to invest in its newberry geothermal project. *Business Wire*. http://www.businesswire.com/news/home/20080313006014/en/Davenport-Power-Announces-Agreement-Renewables-Group-Riverstone. Accessed 15 Aug 2016

Business Wire. (2013). AltaRock Energy announces successful multiple-zone stimulation of well at the Newberry Enhanced Geothermal Systems Demonstration. *Business Wire*. www.businesswire.com/news/home/20130122005635/en/AltaRock-Energy-Announces-Successful-Multiple-Zone-Stimulation-Newberry. Accessed 15 Aug 2016.

Cooper, K. (2010, August 9). Biological evaluation - Threatened, endangered and sensitive plants Newberry Volcano Enhanced Geothermal System Project. Washington, DC: US Department of Interior. Bureau of Land Management. http://www.blm.gov/or/districts/prineville/plans/files/Botany_BiologicalEvaluation.pdf. Accessed 15 Aug 2016.

Davenport Power. (2008). Newberry geothermal power. https://www.wou.edu/las/physci/taylor/newberry/NEWBERRY_GEOTHERMAL_NEWS_2008.pdf. Accessed 5 June 2014.

Ferris, D. (2012). Geothermal company drills into a Volcano. *Forbes*. www.forbes.com/sites/davidferris/2012/10/08/geothermal-company-drills-into-a-volcano. Accessed 6 Oct 2013.

Friedrichs, J. (2007, July 22). Geothermal energy controversy in central Oregon. *NewWest*. http://newwest.net/city/article/geothermal_energy_controversy_in_central_oregon/C509/L509/. Accessed 15 Aug 2016.

Glanz, J. (2009). Deep in bedrock, clean energy and quake fears. *The New York Times*. http://www.nytimes.com/2009/06/24/business/energy-environment/24geotherm.html?pagewanted=all&_r=1&. Accessed 15 Aug 2016.

KTVZ. (2012). Newberry geothermal project receives BLM permit. *KTVZ*. http://www.ktvz.com/news/Newberry-geothermal-project-receives-BLM-permit/-/413192/16670110/-/y00utd/-/index.html. Accessed 15 Aug 2016.

Lecher, C. (2013). Could a volcano power America? *Popular Science*. http://www.popsci.com/science/article/2013-05/volcano-could-power-america. Accessed 15 Aug 2016.

Massachusetts Institute of Technology. (2006). *The future of geothermal energy: Impact of enhanced geothermal systems (EGS) on the United States in the 21st century*. https://www1.eere.energy.gov/geothermal/pdfs/future_geo_energy.pdf. Accessed 15 Aug 2016.

Matteis, J. (2012). Fracking the volcano: Is Oregon project the future of green energy? *The Corvallis Advocate*. http://www.corvallisadvocate.com/2012/1121-fracking-the-volcano/. Accessed 15 Aug 2016.

Molland, J. (2012). Is disturbing a sleeping volcano to get "clean" energy safe? *Care2*. http://www.care2.com/causes/is-disturbing-a-sleeping-volcano-to-get-clean-energy-safe.html. Accessed 15 Aug 2016.

Nelson, G. (2012). Fracking boom lifts hopes for geothermal project at Oregon Volcano. *E&E Publishing*. http://www.eenews.net/stories/1059968049. Accessed 15 Aug 2016.

Newberry Demonstration. (2012). Newberry geothermal project secures sundry permit. *AltaRock Energy*. http://altarockenergy.com/projectupdates/Davenport_Secures_Sundry_Permit.pdf. Accessed 15 Aug 2016.

Newberry Geothermal Energy. (2012, October 20). Pumps are running! *Newberry Geothermal Energy*. https://blog.newberrygeothermal.com/2012/10/20/pumps-are-running/. Accessed 15 Aug 2016.

Oregon Department of Energy. (2008, July 25). Minutes: Energy facility siting council meeting. http://www.oregon.gov/energy/Siting/docs/Minutes/EFSC_7-25-08.pdf. Accessed 15 Aug 2016.

Petty, S. (2010). Geothermal energy protecting the environment – And our future. *AltaRock Energy*. http://altarockenergy.com/projectupdates/WhitePaper.pdf. Accessed 15 Aug 2016.

PLS Environmental, LLC. (2011). Scenic resources inventory and assessment Newberry Volcano EGS demonstration project Deschutes County, Oregon. *US Department of Interior. Bureau of Land Management*. http://www.blm.gov/or/districts/prineville/plans/files/Scenic_Report.pdf. Accessed 16 Aug 2016.

Profita, C. (2012a, January 23). Chemicals from geothermal project concerns doctor in Bend. *Oregon Public Broadcasting*. http://www.opb.org/news/article/chemicals-geothermal-project-concerns-doctor-bend/. Accessed 15 Aug 2016.

Profita, C. (2012b, January 24). Geothermal fractures vs. hydraulic fracking: What's the difference? *Oregon Public Broadcasting*. http://www.opb.org/news/blog/ecotrope/geothermal-fractures-vs-hydraulic-fracking-whats-the-difference/. Accessed 15 Aug 2016.

Profita, C. (2012c, April 5). Newberry Geothermal Project gets a green light. *Oregon Public Broadcasting*. http://www.opb.org/news/blog/ecotrope/newberry-geothermal-project-gets-a--green-light/. Accessed 15 Aug 2015.

Simon, C. (2007). *Alternate energy: Political, economic, and social feasibility*. New York: Rowman and Littlefield Publishers.

ThinkGeoEnergy. (2010). Davenport Power. http://www.thinkgeoenergy.com/tag/davenport-power/. Accessed 1 Aug 2015.

Chapter 6
Oregon Coast Wave Energy

Abstract Wave energy development off the coast of Oregon is still in its infancy. In 2015 a major developer—Ocean Power Technologies—pulled out of its plan to place 100 wave energy buoys off the coast of Reedsport, Oregon. The developer stated that the project had become too expensive and "too complicated" (Schwartz, Wave energy development pulls plug on Oregon project. Oregon Public Broadcasting. http://www.opb.org/news/article/wave-energy-developer-pulls-plug-on-oregon-project/. Accessed 15 Aug 2016, 2015). There are still many areas available for future development with much support from the state government; however, developers will need to demonstrate that "…surrounding ecological and economic resources won't be adversely impacted" (Schwartz 2015). There also are still many critics of wave development, including various fishing interests, environmental groups, and the Oregon Dungeness Crab Commission. With efforts underway to develop the resource and the need for Oregon to meet the 2007 RPS to provide 25% of retail sales of electricity from renewable sources by 2025, the development of wave energy will continue to generate 'green versus green' conflict.

Keywords Wave energy • Green versus green • Technology • Public perception

6.1 Introduction

> *Obviously, these are all issues we'll pay attention to… We're concerned about how new developments in the ocean will affect the fish population.* (Shawn Miller, lobbyist for the Coastal Conservation Association (Griegerich 2012))
>
> *Let's get this stuff in the water and find out what it has to offer…Consumers want green power, and this is an option.* (Philip D. Moeller, Member of the Federal Energy Regulatory Commission (Yardley 2007))

Wave energy technology has burgeoned in recent years with the deployment of test buoys and demonstration sites off of the coastlines of Australia, France, the United Kingdom, and the United States. Yet, communities on the edge of these

J.C. Pierce, B.S. Steel, *Prospects for Alternative Energy Development in the U.S. West*, Environmental Challenges and Solutions 4,
DOI 10.1007/978-3-319-53414-5_6

developments have had various degrees of success in engaging with project developers and making informed decisions about the role that wave energy should play in their future. In fact, public perception of wave energy varies widely across geographies, cultures, and demographics. What do communities think about emerging wave energy technology? And, what is really going on between wave energy developers and the communities in which they hope to build sites?

Public perception is generally positive for renewable energy development in principle, but declines when it is proposed in a specific place. Installation of the technology, especially in geographic proximity to the proposed site, tends to result in an increase in public support (Firestone et al. 2009). Research on public opinion of wave energy development in Oregon found that, in general, men have positive attitudes towards wave energy development, whereas the majority of women feel that they do not have enough information to form an opinion (Stefanovich 2010). The same study also found that the level of education of respondents was positively correlated with supportive attitudes towards wave energy development.

In the European context, researchers have found that stakeholders, ranging from organizations to grassroots members, were satisfied with the level of consultation in wave energy development projects (Bailey et al. 2011). However, stakeholders raised concerns about the project developers over-stating potential benefits, especially economic benefits, in order to gain community support and buy-in; the stakeholders also complained that the information provided by companies was often long and technical and that there was also some disparity in who was invited to and participated in meetings. For example, in the study, surfers and fisherman were more likely to be invited and participate than were divers or recreational boaters. The consultation process under study was for a test site. Community members and other stakeholders wanted to ensure that the approval of a test facility did not necessarily mean that subsequently they would approve a commercial site.

Communities opposed to wave energy typically cite negative aesthetic, noise, or wildlife impacts (Bailey et al. 2011). While ave energy is generally less visually invasive than offshore wind development, for example, it is still visible in most cases and therefore is of particular concern to tourist-dependent coastal communities. In large part, local citizens maintain that there are a number of unanswered questions. What types of business practices and regulations will be necessary to protect the ocean and other important economic drivers, such as fishing and recreation? What are the potential impacts of wave energy development? How will the community be involved in the decision-making process? Are there opportunities to engage in community financing options? And, in the future, how might ocean acidification, tsunamis, or debris and damaged equipment be dealt with?

Occupation, reliance on the ocean, and connection to place also play important roles in perception. Crabbers, fishermen, state agencies, wave energy developers, and scientists maintain different views on development and are likely to benefit or bear costs in very different ways (Stefanovich 2010). More research is needed to discern how perception varies across occupational groups and at different stages of wave energy development. Overall, symbolic interpretations of wave energy play a critical role in understanding support and opposition for local development (Bailey et al. 2011; Stefanovich 2010).

Wave energy technology is still largely in its infancy, and wave energy companies need to be cautious of overstating economic opportunities. Economic models that include realistic multipliers and job creation numbers will be crucial to the long-term success of wave energy development. Public perception of wave energy technology will be an area of continued research as more projects move from concept to reality. Wave energy developers are interested in learning how to best work with communities in order to ensure the successful development of projects. At the same time, rural coastal communities are in need of jobs and economic development opportunities, making wave energy development all the more enticing.

6.2 Case Study Project

The development of the wave energy industry has been on the radar of Oregon political leaders since 2005. Oregon is viewed as being favorably positioned to be a major player in the wave energy industry. Over 300 miles of coastline, powerful waves, steady winds, and existing infrastructure may make wave energy an ideal source of power when compared to other resources. According to the Oregon Wave Energy Trust (OWET), wave power generation would be a highly predictable and reliable form of energy with the potential for low costs of grid integration (Oregon Wave Energy Trust 2013). Oregon is thus slated to be the first U.S. state to test and potentially develop grid-scale wave energy.

6.2.1 Testing

In 2008, the U.S. Department of Energy created the Northwest Marine Renewable Energy Center (NNMREC), which is a partnership between the University of Washington and Oregon State University. While OSU is focusing on wave energy development off the Oregon coast, the UW is focusing on tidal energy in Puget Sound. Although wave energy has yet to be developed and implemented in the Pacific Northwest, laboratory testing is taking place at OSU's O.H. Hinsdale Wave Research Laboratory and the Energy Systems lab in the Wallace Energy Systems and Renewables Facility (WESRF). Various testing sites off the Oregon coast have been identified by NNMREC, including a site of about one square mile that lies about two miles northwest of Newport, Oregon.

Several companies have come to Oregon to test various technologies, including the company M3 Wave, which plans to test a deep-water variant of their DMP/APEX technology. These devices turn ocean waves into "alternating expansion and contraction cycles of an air-filled system which, in turn, drives an air turbine connected to an electrical generator. As [a] wave passes over one chamber of the device,

Fig. 6.1 Example wave energy buoy (National Renewable Energy Laboratory)

the pressure increases, forcing the air to move through the air turbine and into the second chamber. The moving air forces cause the turbine to spin, driving the electrical generator to produce electricity" (Tidal Energy Today 2015). According to the developer, the device has no exposed mechanical components and only one moving part—the generator. The first test device, deployed off of Astoria, is a small-scale test model that is 30 feet long and 8 feet wide.

Other wave devices to be tested off the Oregon coast include the traditional buoy devices, like the PowerBuoy (PB 150), which is tethered to the ocean floor and floats on the surface and uses wave action to power an encased generator. Electricity generated by the buoy is then transmitted to the shore and grid via underwater power cables. According to Ocean Power Technologies, the company developing the devices, "The majority of the PB 150 structure is hidden from view underwater—its height above the waterline is only 30 feet. When deployed at a typical site 2 ½–3 miles out to sea, the size of the PowerBuoys when viewed from shore would be [equivalent to] approximately 1.6 mm when viewed from arm's length" (Oregon Power Technology 2015). An example of a wave buoy can be found in Fig. 6.1.

6.2.2 Development and the Territorial Sea Plan

Even with the promise of a clean, renewable energy source, however, wave energy development has faced significant opposition in the State of Oregon. This has been largely due to the limited information available to stakeholders, environmental concerns, high initial costs, and the limited number of wave energy installations compared to other sources of energy. The desire for a clean power source nevertheless has largely driven the development of wave energy on the Oregon coast and was a major influence in the development of the Territorial Sea Plan. The plan is an attempt to balance the multiple uses of Oregon's territorial waters while limiting the negative impacts on the environment and on a variety of stakeholders. Heralded as a great compromise, the Territorial Sea Plan is intended to shape the development of wave energy projects without severely limiting other ocean uses. As former Oregon Governor John Kitzhaber remarked (News Lincoln County 2012):

> Oregon has long been a leader in renewable energy development, and energy issues will have the single greatest impact on Oregon in the coming decade. This balanced proposal shows Oregon can thoughtfully support this emerging and promising industry while protecting our coastal communities' quality of life, our commercial and recreational fisheries, and a coastline that all Oregonians treasure.

Over the years, numerous locations have been discussed as having the potential to support a wave energy installation. The debate was finally settled in January 2013 when the Land Conservation and Development Commission approved an enhanced Territorial Sea Plan for Oregon (see the Box 6.1 for the highlights of the plan).

The new plan included policies aimed at governing the development of wave energy along the Oregon coast and named four sites as suitable for wave energy development: Reedsport, Camp Rilea, Nestucca River, and Lakeside (see Table 6.1 for details of Oregon wave energy projects). Of these sites, three have potential projects. The fourth site, Lakeside, thus far is not being explored for a project, but there may be some effort in the future.

Box 6.1 Highlights of the Territorial Sea Plan for Oregon

- Four sites designated as suitable for wave energy development; two for near-shore technology and two for deepwater technology.
- Sites combine to make up 2%, or 22 square miles, of Oregon's territorial seas.
- Up to 3% of coastal waters may be made available to future wave energy developers.
- Cap of 3%, or 36 square miles, of Oregon's territorial waters can be utilized for the development of wave energy.

Table 6.1 Timeline of wave energy projects

Date	Action
October 2006	The Reedsport Wave Energy Project was designated as an Oregon Solutions project by Governor Kulongoski (Oregon Solutions 2013).
2007	Wave energy was selected as an economic innovation focus by the Oregon Innovation Council. The Oregon Wave Energy Trust (OWET) was then established to oversee "the responsible development of wave energy projects in Oregon."
January 2007	OPT files application for preliminary permits to look into the feasibility of wave power stations at two Oregon locations: Coos Bay and Newport.
February 2007	Preliminary permit issued by U.S. Federal Energy Regulatory Commission (FERC) to Ocean Power Technologies (OPT) to "the feasibility of a location off the coast of Reedsport, Oregon for the proposed construction and operation of a wave power station with an anticipated maximum rated output of 50 MW, of which up to the first 5 MW will be a demonstration wave power station." OPT signs a cooperative agreement with Pacific Northwest Generating Cooperative (PNGC) to aid in the development of a wave station.
June 2007	U.S. Navy awards OPT a $1.7 million contract to utilize their PowerBuoy.
July 2007	OPT files Pre-Application document and Notice of Intent with FERC to show intent to seek a license for a wave park in Reedsport, OR. It is believed that OPT is the first wave power company to file these documents.
August 2010	Ocean Power Technologies (OPT) signed agreements with 11 federal and state agencies and three non-governmental stakeholders to support the development of the next phase of the Reedsport Wave Energy Project – a 10-PowerBuoy, grid connected wave energy station with a capacity of 1.5 MW. The agreement is a key step towards the granting of a license by FERC, which would be the first such license to be issued for a commercial-scale wave power project in the U.S. (Oregon Solutions 2013).
November 2010	Aquamarine Power, a Scottish wave energy company, invests in Oregon wave energy and opens an office in Newport with the intention of exploring wave energy projects in the state.
November 2011	Aquamarine Power closes its Newport office and leaves the state citing a lack of certainty over the future of wave energy in Oregon due to the Territorial Sea Plan.
December 2011	The Electric Power Research Institute published a study detailing Oregon's total available wave energy. It estimated that the inner shelf alone is capable of producing equal to 143 terawatt-hours per year (TWh/yr), or 143 billion kilowatt-hours per year (KWh/yr).
December 2011	DEQ report released evaluating the use of Reedsport as a wave energy project site. The findings generate a level of controversy due to potential impacts on water quality, environmental degradation, and fishing impacts.
May 2012	In response to wave energy development, Clatsop County Board of Commissioners adopt Statewide Planning Goal 19. This goal is intending to protect existing ocean users from the impacts of wave energy by tightening the permitting and licensing process at the county level.
June 2012	OPT announces the successful testing of a utility scale PowerBuoy, the PB150, which is supposed to be deployed off the coast of Reedsport.

(continued)

Table 6.1 (continued)

Date	Action
August 2012	FERC issues permit to OPT to construct a "grid-connected" wave power farm off the coast of Reedsport, OR. This 1.5 megawatt power station would consist of 10 buoys and would provide enough energy to power 1,000 homes.
September 2012	Pacific Marine Energy Center receives first round of funding. Payment consists of $4 million from the United States Department of Energy and a non-federal cost match.
November 2012	Waldport is taken off the list of primary sites for wave energy development.
January 2013	Newport selected as the site for Oregon State University's Northwest National Marine Renewable Energy Center (NNMREC).
January 2013	The Territorial Sea Plan (TSP) is approved and names four locations as Renewable Energy Facility Suitability Study Sites (REFSSAs). The TSP also includes policies that are designed to govern and set the conditions for ocean energy projects in the state.
August 2013	Wave energy project off the coast of Reedsport stalls after OPT violated its license by failing to submit necessary plans and assessments for the project.
March 2014	OPT surrendered its permit to develop a wave energy project off of Reedsport, siting high costs and significant constraints to development.
2015	OPT pulled out in 2015

6.3 Reedsport Site

The Reedsport site off the southern Oregon coast is viewed as having great potential for wave power generation. In October 2006, Governor Kulongoski designated the Reedsport Wave Energy project an Oregon Solutions Project in an effort to foster collaboration and stakeholder involvement. Some of the factors that influenced this decision included existing infrastructure, community support, support services, and ocean depth. The Reedsport project was also seen as an opportunity to work through the process of planning and implementing a wave energy project so that future projects could be more streamlined.

Ocean Power Technologies (OPT) was selected as the developer of the Reedsport wave energy park project, which would be the first commercial wave energy facility on the West Coast. The company was granted the first permit from the U.S. Federal Energy Regulatory Commission (FERC) for wave energy development. (To date, FERC has granted only a limited number of permits for wave energy developers; however, many advocates believe that the agency is still trying to determine how to best deal with licensing proposed projects.) The company was also awarded over $4.6 million in grants from the U.S. Department of Energy.

In August 2012, FERC issued a permit to OPT to construct a "grid-connected" wave power farm off the coast of Reedsport. The estimated 1.5 megawatt power station would consist of 10 buoys and would provide enough energy to power 1,000 homes (Oregon Power Technology 2015). The process would be slow moving, however, due to the involvement of numerous stakeholders wanting to monitor every phase of the project and ensure its careful execution. As Ron Yockim, from RPS Consulting, commented (Manning 2009), "This contract that we're getting prepared to issue is one that's probably critical. It allows us to do not only an energy analysis, that says what the size of the wave is, what the energy components are, but it's also a critical force in terms of looking at what the design should be of this project."

The Reedsport wave energy project had unequivocal support from high ranking political figures and community members. Governor Kulongoski had long promoted the development of wave energy in Reedsport, citing the potential monetary and environmental benefits wave energy had to offer. To ensure the sustainability of Oregon's energy market, Kulongoski advocated the advancement of a "green" power solution. The governor's position was echoed by FERC Commissioner Philip Moeller, who stated that wave energy was a viable green power option that needed to be more aggressively developed. Approving the OPT permits for the Reedsport site would be the first step in this direction. Federal Energy Regulatory Commissioner Philipp Moeller commented (Yardley 2007), "Let's get this stuff in the water and find out what it has to offer. Consumers want green power, and this is an option."

Despite the promise of wave energy, the Reedsport project faced considerable backlash from a number of stakeholders. Local residents and area fishermen expressed concerns regarding decreased water quality, environmental degradation, and wave energy buoys impeding access to high quality fishing areas. The proposed project site would have been placed in what were viewed as some of the best crab fishing areas in the state, which prompted local fishermen to strongly object to the project. In response to these concerns, OPT took the stance that the location of the wave energy project was non-negotiable. However, despite the initial reluctance of OPT to move the site, area fishermen felt it was in their best interests to sign a settlement agreement with OPT because such an agreement would be binding under the Federal Power Act (FPA) (Husing 2011). Essentially any agreement made between OPT and any party would be enforceable under federal law.

Another issue raised in opposition was that of the power granted to OPT under the FPA. Section 4(e) of the FPA stated that, for the duration of the permitting process, FERC must grant "equal consideration" to the development of wave energy, while also considering the protection of environmental, cultural, and recreational resources. This meant that fishermen would receive no protection from the impacts of the wave energy project and that OPT would be granted exclusive rights to utilize that part of the ocean, resulting in the loss of prime fishing grounds. Any uses within the designated project area would be solely based on the discretion of OPT. In response, fishermen asked OPT to move the project site to the west, which would limit the impact on the fishing industry. Doing so, according to OPT, would diminish the power-generating capacity of the project, as the current location was seen as the ideal location in terms of generating capacity.

Unknown impacts were also at the heart of the controversy. Many stakeholders have compared wave energy to the building of dams. Oregon Salmon Commissioner Nancy Fitzpatrick observed (Yardley 2007): "Is it going to impact us? Going way back to the dams, we find out later that of course, yes, it affected salmon and migration. So we don't want to be stuck in a situation like that with wave energy." Though wave energy may have minimal impacts, at least in terms of our current knowledge, the concern was that the technology is too new for sufficient understanding of long-term ramifications. Other potential concerns included increased turbidity, decreased water quality, and the buildup of toxic substances in navigable waters.

Under the DEQ permitting process, OPT had to "provide DEQ with reasonable assurance that the construction and operation of the proposed project will not contribute to violations of water quality criteria" (Oregon Department of Environmental Quality 2011). In a 2011 DEQ report on the Reedsport OPT project, various potential concerns were noted; however, DEQ stated that it had "reasonable assurance" that OPT would comply with any and all provisions and statutes that fell under the scope of the project. Furthermore, under the Territorial Sea Plan, OPT could not sell or transfer the project site to any other company.

In 2013, however, the project stalled, in part because the company violated their permit and failed to meet deadlines for deployment, causing FERC to intervene (Hunt and Cardwell 2014). OPT surrendered its permit to develop a wave energy project, and, in 2015, pulled out of its plan to place 100 wave energy buoys off the coast of Reedsport. The developer stated that the project had become too expensive and "too complicated" (Schwartz 2015). The site will transition to a Resources and Uses Conservation Area (Benham 2013).

6.4 Warrenton (Camp Rilea) Site

In January 2013, Camp Rilea near Warrenton was named one of four potential sites for the development of renewable wave energy. Support for the placement of a wave energy station on the north coast of Oregon at Camp Rilea has been high for the following reasons: (a) existing infrastructure, (b) lower military energy costs, (c) value of renewable energy, (d) economic development, and (e) job creation (Oregon Wave Energy Trust 2015). Numerous stakeholders, including community members, businesses, and Clatsop Community College, have voiced their support for the project and believe wave energy will be a major economic component in the future. This sentiment was boosted by a 2012 assessment by the Oregon National Guard, which highlighted what it believed were the benefits of Camp Rilea over other sites (Oregon Army National Guard 2013):

- The Oregon National Guard has engineers, planners, and compliance staff on-site which would reduce the need for an outside agency to oversee the project.
- Placement of the site would help the Army achieve total Net Zero by 2020.
- As a national security site and a disaster response site, Camp Rilea needs to have multiple power sources available to maintain "resiliency."
- Wave energy projects would create jobs, boost the economy of the North Coast Area, and enhance the response of emergency services for Camp Rilea.

- The potential project area is the site of military training exercises and is shut down often. This would minimize impacts when compared to other areas.
- Existing electrical infrastructure, including submarine cables and an electrical distribution system close to the shore, are other benefits of the site.

The location and existing infrastructure are widely cited as the major reasons for building a wave energy station at Camp Rilea. The Oregon Army National Guard has been a vocal supporter of the project and of the cost savings of renewable energy. It is estimated that a wave energy site would save the military approximately $2 million per year, while helping them reach their Net Zero goals (Oregon National Guard 2013). Other organizations, including Clatsop Community College (CCC) and Clatsop County, believe that wave energy provides an opportunity for economic growth and job creation. CCC has taken steps to rent its research vessel to companies for the deployment of wave energy devices. Dean of Workforce Education at CCC, Kristen Wilken, views wave energy as an opportunity to connect wave energy companies with students majoring in maritime science. Clatsop County leaders believe wave energy will provide an economic benefit to the area through increased employment opportunities, decreased unemployment rates, and increased development of business parks around Camp Rilea (Oregon Wave Energy Trust 2015).

Critics of the project have expressed concerns in regard to the resilience of the companies involved in the wave energy project. Lincoln County Commissioner Terry Thompson generally supports the idea of wave energy, but is apprehensive due to the high costs involved, the durability of the equipment, unknowns about the technology, environmental impacts, and the large amount of grants paying for these projects. The large amount of grant-based funding and money from outside sources calls into question how viable these companies will be in the long run. Thompson also warns that every community is different and that a one-size-fits-all approach will not work with wave energy development. The key, he says, is to "make sure communication is happening back and forth with that community" (Shorack 2013). Peter Huhtala, chairman of the Board of Clatsop County Commissioners, also believes that communication is vital, given the relative risks of wave energy and the limited information available regarding environmental impacts: residents place a high value on the ocean as a resource and are very protective of their jobs, food, and enjoyment. In order to gain the buy-in of area residents, companies need to communicate effectively and build trust. As Peter Huhtala, Chairman of the Board of Clatsop County Commissioners commented (Shorack 2013), "Oregon manages the public trust of its territorial sea with a fundamental preference for protecting marine life and sustainable uses. As we add another sustainable use, we want to get it right."

Local fishermen have been vocal in their criticism of the proposed project, largely due to the potential impact of the project on the fishing industry. However, limited information is available regarding the actual impact(s), which has raised tensions. The local fishing industry is composed of a large fleet of small boats that stay close to port so they can get quickly get back if weather conditions deteriorate. John Corbin, commissioner of the Oregon Dungeness Crab Commission, has argued that the safety of local fisherman could be at risk if they are forced to travel farther out to fish.

Another concern is the intrusion of the proposed project study area into prime crab fishing grounds. Like the south coast Reedsport site, a significant portion of the proposed north coast site (11 square miles) is viewed as having some of the best crab fishing in the state. According to North Coast Commercial Fishing, a significant portion of the study area lies in the top 20% for crab fishing, while a smaller portion is in the top 10% (Shorack 2013). Scott McMullen, a representative with North Coast Commercial Fishing who worked on the Territorial Sea Plan as a member of Ocean Policy Advisory Council (OPAC), believes a smaller study area is needed, due to the high value placed on the area by fishermen. In response to these concerns, OPAC recommended that the study area be reduced in size to three square miles. However, the Land Conservation and Development Commission (LCDC) disregarded OPAC's recommendations and instead endorsed the original 11 square mile study area. Scott McMullen, from North Coast Commercial Fishing commented (Shorack 2013), "The commission's decision didn't protect those areas as well as OPAC's (recommendation). The staff report is less protective of the fishing industry."

Concerns from Clatsop County echoed many of the concerns heard at other sites: environmental degradation and lack of access to prime fishing grounds topped the list. Clatsop County officials hope to work with State officials to adopt a process where input from both State and County leaders is heard. With many local stakeholders maintaining the stance that fishing and wave energy are incompatible uses, further dialogue now may help mitigate concerns and allow the state and county to work towards an amicable solution. For example, as Peter Huhtala, Clatsop County Commissioner, observed (Shorack 2013), "In coming years new uses will compete for space in the ocean. One emerging ocean industry will create electricity from waves, wind and tidal forces. As we nurture this renewable ocean energy industry we must respect and protect existing ocean users."

6.5 Nestucca River Site

The Nestucca River site near the north-central coast was chosen due to its geographic benefits and existing infrastructure. The ocean floor at this location is flat and sandy, which is ideal for a wave energy installation. Supporters also noted that a power substation capable of handling the increased load is only 500 yards away; that substation location would save time and reduce development costs. The Oregon Wave Energy Trust (OWET) publically supported this site and also noted that any wave energy devices installed at this location would be subsurface or minimally visible (Wrabek 2013).

Yet, critics believe the development of wave energy at this location would adversely impact the fishing industry and aquatic life (*News Lincoln County* 2012). The proposed location of the study site is south of the mouth of the Nestucca River in prime fishing waters. Area fishermen have voiced the concerns that wave energy development would limit their access to quality fishing areas and impede migrating salmon and steelhead. Many jobs, they argue, would be at risk, as area fishermen would need to find new fishing locations farther away from shore. Other critics cite the inherent risks of placing steel devices or machinery into the ocean. These devices would require significant maintenance at a high price tag (*News Lincoln County* 2012). If not maintained properly, these devices could pose a threat to recreational users, aquatic life, and the fishing industry.

The Department of Land Conservation and Development (DLCD) and the Land Conservation Development Commission (LCDC) recommended the Nestucca River location as a Renewable Energy Facility Suitability Study Site (REFSSA), despite significant opposition (Wrabek 2013). Both OPAC and the Territorial Sea Plan Advisory Committee expressed concern over the location of the site and recommended that the site be removed from the REFSSA list. Concerned citizens, area fishermen, and recreationists also expressed their concerns to the DLCD and LCDC, but saw their efforts to have the site removed from consideration rebuffed.

During public hearings on the matter, several organizations and interest groups expressed unease over the location of the proposed wave energy site. With the support of area fishermen, OPAC recommended removing the Nestucca River site from the list of possible sites due to its proximity to prime fishing areas. Community members also urged the commission to remove the site from consideration because it is highly valued by the community. The LDCD, however, voted 5-1 to put the Nestucca River location back on the list of REFSSAs. The vote was met with immediate criticism from multiple directions. Nehalem Mayor Shirley Kalkhoven announced to the Tillamook County Board of Commissioners that David Yamamoto of Pacific City and Charles Ciecko of Neskowin would be appealing the decision of the DLCD. The decision to appeal, according to Kalkhoven, was based on the DLCD allegedly ignoring convincing testimony from two advisory committees, as well other stakeholders, to remove Nestucca River from the list of proposed sites (Wrabek 2013). 6.6 Lessons Learned

Wave energy development off the coast of Oregon is still in its infancy. There are still many areas available for future development with much support from the state government; however, developers will need to demonstrate that "surrounding ecological and economic resources won't be adversely impacted" (Schwartz 2015). There are also still many critics of wave development, including various fishing interests, environmental groups, and the Oregon Dungeness Crab Commission. With efforts still underway to develop the resource and with the need for Oregon to meet the 2007 RPS to provide 25% of retail sales of electricity from renewable sources by 2025, the development of wave energy will continue to generate "green versus green" conflict.

6.6 Case Studies: Conclusions

The case studies illustrate the complexity of environmental tradeoffs between concern for the local environment, economy, and other considerations versus the need to generate electricity from non-carbon based energy resources. They also illustrate the need for those who hope to successfully develop renewable energy projects to involve and to establish trust with potentially affected local stakeholders.

As we saw in this chapter, people whose livelihoods depend on ocean resources—such as fishers and crabbers—may feel threatened by new ocean uses that may harm their traditional occupations. There may be community aesthetic concerns about how wave energy may affect tourism and scenic beauty or, as in the case involving wind power (Chap. 4), scenic beauty and property values. Likewise, there may also be environmental concerns, such as the impact of new structures on the ocean floor or, as we saw in Chap. 3, about the siting of a solar plant where it could affect wildlife habitat. There may even be fears about the effects of AET on the earth itself—for example, by drilling into a volcano, as we saw in Chap. 5. Supporters of solar, wind, geothermal, and wave energy, on the other hand, may see the potential economic benefits of AET projects, especially for economically distressed rural communities. Their support may also stem from the desire to combat the effects of global climate change or to decrease reliance on fossil-fuels in the face of possible future scarcity.

Both support for and opposition to AETs may be linked to many different variables, including individuals' core environmental values, knowledge levels, sources of information, demographic factors, political leanings, and other considerations. Although these four case studies provide valuable insights, examples, perspectives, and lessons learned, case studies are anecdotal by nature and are necessarily limited. Hence, more information is needed.

The next section works to explore those linkages through opinion surveys of the general publics of California, Oregon, Washington, and Idaho in order to identify empirically distinctive combinations of environmental and AET opinions, and then to describe how those types differ across such attributes as values, attitudes, demographic characteristics, scientific knowledge and geographic location. To place the survey results in context, Chaps. 7 and 8 also provide overviews of key concepts that may be helpful in understanding the connection between environmental values and the choices individuals make about support for or opposition to alternative energy technologies.

References

Bailey, I., West, J., & Whitehead, I. (2011). Out of sight but not out of mind? Public perceptions of wave energy. *Journal of Environmental Policy & Planning, 13*(2), 139–157.

Benham, S. (2013). Diving into ocean power: State approves plan for wave energy sites. *Katu News*. http://www.katu.com/politics/Diving-into-Ocean-power-State-approves-plan-for-wave-energy-sites-188444881.html. Accessed 15 Aug 2016.

Firestone, J., Kempton, W., & Krueger, A. (2009). Public acceptance of offshore wind power projects in the USA. *Wind Energy, 12*, 182–202.
Griegerich, A. (2012, July 26.). Wave energy controversy swells in Oregon. *Sustainable Business Oregon.*. http://www.bizjournals.com/portland/blog/sbo/2012/07/wave-energy- controversy-swells-in-oregon.html. Accessed 15 Aug 2016.
Hunt, J., & Cardwell, D. (2014, April 12). Experimental efforts to harvest the ocean's power face cost setbacks. *The New York Times*, April 12. http://www.nytimes.com/2014/04/28/business/energy-environment/experimental-efforts-to-harvest-the-oceans-power-face-cost-setbacks.html?_r=0. Accessed 19 Dec 2015.
Husing, O. (2011). The challenges and opportunities of renewable energy development on the Oregon Coast. *Oregon Coastal Notes*. http://www.oczma.org/pdfs/Coastal%20Renewable%20Energy%20Final%204-11_1.pdf. Accessed 15 Dec 2015.
Manning, R. (2009, July 17). Douglas County wave project rolls forward. *Oregon Public Broadcasting*. http://www.opb.org/news/article/douglas-county-wave-project-rolls-forward/. Accessed 15 Aug 2016.
News Lincoln County (2012). State decides which offshore fishing grounds will be shut down by wave energy devices – Governor Kitzhaber reacts to LCDC decisions. *News Lincoln County.*, http://www.newslincolncounty.com/archives/72963. Accessed 15 Aug 2016.
Ocean Power Technology (2015). OPT PB 150 PowerBuoy. http://www.oceanpowertech.com/PDF/PB150_08292012_GFL.pdf. Accessed 15 Aug 2016.
Oregon Army National Guard. (2013). Net zero newsletter. http://www.asaie.army.mil/Public/Partnerships/doc/Issue_1_Jan_13.pdf. Accessed 15 Aug 2016.
Oregon Department of Environmental Quality (2011). *Reedsport OPT wave park evaluation and findings report*. http://www.deq.state.or.us/wq/sec401cert/docs/hydropower/reedsportOPT/EvalFindingsReport.pdf. Accessed 15 Aug 2016.
Oregon Solutions (2013). Reedsport wave energy project. *Reedsport Wave Energy Project*. http://orsolutions.org/osproject/reedsport-wave-energy/. Accessed 15 Aug 2016.
Oregon Wave Energy Trust (OWET). (2013). Ocean energy. http://oregonwave.org/information/ocean-energy/. Accessed 10 July 2014.
Oregon Wave Energy Trust (OWET). (2015). Wave energy industry update: A northwest US perspective. http://oregonwave.org/oceanic/wp-content/uploads/2015/10/Northwest-Wave-Energy-Industry-Update-FINAL.pdf. Accessed 15 Aug 2016.
Schwartz, D. (2015, February 18). Wave energy development pulls plug on Oregon project. *Oregon Public Broadcasting*. http://www.opb.org/news/article/wave-energy-developer-pulls-plug-on-oregon-project/. Accessed 15 Aug 2016.
Shorack, T. (2013, September 26). Groups set stage for wave energy collaboration. *The Daily Astorian*. http://www.opb.org/news/article/groups-set-stage-for-wave-energy-collaboration/. Accessed 15 Aug 2016.
Stefanovich, M. (2010). *Determinants of public opinion on renewable energy: The case of wave energy development in Oregon* (PhD Dissertation). Oregon State University. http://ir.library.oregonstate.edu/xmlui/handle/1957/18625. Accessed 15 Aug 2016.
Tidal Energy Today. (2015). *M3 wave's DMP wave energy converter*. http://tidalenergytoday.com/2015/08/14/video-m3-waves-dmp-wave-energy-converter/
Wrabek, J. (2013). LCDC votes to keep Pacific city on wave energy development list. *The Daily Astorian*, January 31. http://www.dailyastorian.com/20130131/lcdc-votes-to-keep-pacific--city-on-wave-energy-development-list. Accessed 15 Aug 2016.
Yardley, W. (2007, December 8). Efforts to harvest ocean's energy open new debate front. *The New York Times*. http://www.nytimes.com/2007/12/08/us/08waves.html?r=0. Accessed 15 Aug 2016.

Part III
Surveys

Chapter 7
The Tradeoff Terrain

Abstract This chapter describes state level survey data that will be used in the forthcoming chapters, including the survey structure, survey methodology, and indicators used to assess environmental values and tradeoffs between renewable energy development and environmental considerations. This chapter also examines the impact of environmental values as measured by the New Ecological Paradigm—controlling for various socio-demographic characteristics—on hypothetical tradeoffs between solar, windpower, geothermal, biomass and wave technologies on the one hand and potential environmental damage on the other hand.

Keywords Environmental-energy tradeoffs • New Ecological Paradigm • Postmaterialist values

7.1 Introduction

> *There's this theory that you have to pick one: economics or environmental performance. That's nonsense. Innovation is the way you can have both.* (Mark Vachon, Vice President of General Electric (Grover 2012)).
>
> *When it comes to clean energy, it's always been about the trade-off.* (Columnist Steve Sebelius, *Las Vegas Review* (Sebelius 2013))

Conflicts over important issues of public policy, such as the appropriate modes of energy production, often are structured by the political and cultural context within which they take place (Bell et al. 2005; Ellis et al. 2007; Haggett 2011; Kraft and Clary 1991; Leiserowitz et al. 2011; Phadke 2010; Pierpont 2009). Such political and cultural contexts differ significantly, not only between the United States and other countries, but also among various regions of the United States (Leiserowitz et al. 2011; Petrova 2013). Moreover, issues embedded in the siting of alternative energy technologies have major implications for a broad range of personal and political values (Bidwell 2013; Gross 2007; Pasqualetti 2011; Pierpont 2009; Warren et al. 2005; Wolsink 2007, 2010). Thus, to the degree that significant individual beliefs and values connect substantively with the values implied in AET siting proposals, one also would expect individuals' responses to those proposals to vary significantly in accordance with differences in those values.

J.C. Pierce, B.S. Steel, *Prospects for Alternative Energy Development in the U.S. West*, Environmental Challenges and Solutions 4,
DOI 10.1007/978-3-319-53414-5_7

This section of the book reports the results of public surveys in Washington, Oregon, California, and Idaho. This chapter describes the survey structure, and it provides an overview of public attitudes about some of the central elements of conflict over alternative energy issues, how the publics of the four states differ on those attitudes, and the degree to which those attitudes are rooted in fundamental views of the relationships of humans and the environment. We identify public attitudes in response to a set of tradeoff questions regarding the installation of an alternative energy technology (AET) on the one hand, and the potential negative environmental effects of that technology on the other hand. In particular, the focus is on identifying the extent to which *opposition* to AETs designed to minimize global environmental damage from carbon-based energy modes appears to stem from perceptions of negative localized environmental effects from those AETs, even though the AETs themselves are widely seen as environmental initiatives

7.2 Survey Design and Structure

State-level survey data used in the empirical chapters throughout the section are described below, including the survey structure, survey methodology, and indicators used to assess *environmental values*. This chapter also assesses the connections between those environmental values and the choices individuals make about support for alternative energy technologies (AETs) *given certain environmental costs*—i.e., *tradeoffs*. And, it examines the impact of environmental values as measured by the New Ecological Paradigm—controlling for various socio-demographic characteristics—on hypothetical tradeoffs between solar, windpower, geothermal, biomass, and wave technologies on the one hand and potential environmental damage on the other hand.

Four types of respondents will be identified: *globalists* (people with pro-environmental values who support AETs even with local environmental costs); *localists* (people with environmental values who reject AETs when there are local environmental costs); *rationalists* (people who support AETs even though they do *not* hold strong environmental values); and, *traditionalists* (people who do *not* express strong environmental values and who do not support the siting of AET). Much of the focus in the following chapters is on the differences that exist between globalists and localists. The remaining chapters in this section focus on a series of possible explanations for support for AETs (values, ideology, climate change, energy policy beliefs, knowledge, and information sources) when controlling for environmental values, as well as the role of these possible explanations in differentiating between globalists and localists who share pro-environmental values.

From an analysis of surveys of the publics in California, Oregon, Washington, and Idaho, it is possible to gather a sense of citizens' orientations toward the environment and toward the installation of AETs designed to protect the global environment, particularly when the installation of those technologies also may have the potential to harm the local environment. Moreover, the surveys inquire into an individual's attitudes and values that may help account for why, for example, one may exhibit strong environmental orientations, but oppose the installation of AETs that have been justified in terms of reducing carbon emissions leading to global warming. Or, on the contrary, why some people express strong support for the installation of those AETs even though they may be very weak in pro-environmental orientations. Thus, one can begin to approach answers to such questions as whether those who take various positions on the tradeoff questions are more or less (a) environmental in fundamental values, (b) convinced of the climate change effects of traditional technologies, (c) liberal in their political orientations, (d) knowledgeable about energy and the environment, (e) "post-material" in their fundamental political values, or (f) reliant on particular sources of information. The answers to those questions may well structure the future of AETs, as well as the strategies best employed to achieve that future.

In order to address these questions, a mailed survey was sent to random samples of over 1400 households *each* in California, Idaho, Oregon, and Washington in the fall of 2012 and winter of 2013 (see Appendix B). The samples were provided by a commercial research company that has exhaustive databases of households comprising telephone directories, state departments of motor vehicles, and other household information sources. Dillman's (2010) *Tailored Design Method* was used in questionnaire design and its implementation. Each contacted household received the following request for participation: "If available, we would prefer the person, 18 years old or older, who most recently celebrated a birthday to complete the survey." Three waves of the mail surveys were distributed to each sample, followed by a final telephone reminder if necessary. Respondents were also provided an Internet-based survey as a response option. The response rates and the potential response bias statistics are shown in Table 7.1.

The response rates varied only marginally among the four states, with the highest percentage in Oregon (51.5%) and the lowest in Idaho (46.6%). These small differences may reflect the relative salience of the AET installation issues in the two states. Indeed, many more visible public controversies have emerged in Oregon than in Idaho, and the same is likely true for Washington and California. In California, the responding sample was generally representative of the known population attributes, with the exception of education; the sample reflected a higher level of education than for the California population as a whole: 49.1% compared to 36.7%, respectively. Gender, income, and age characteristics in the sample mirrored the population more closely than did the education variable.

The Idaho sample likewise was very similar to the known population attributes for age, income, and gender. The Oregon population and sample data also were very similar to each other, with perhaps the lone exception of the income variable—the most frequently chosen income category was higher in the sample responses than was the actual median household income category in the state. On the other hand, the education levels of the two groups were very close to each other. In regard to education level, the potential sample bias patterns in Washington were more similar to those in California than to those in Oregon and Idaho. That is, about 10% fewer

Table 7.1 Survey response rates and bias data survey response rates

State	Date	Sample size	Responses	Response rate
California	Fall, 2012	1433	693	48.3%
Idaho	Winter, 2013	1413	687	46.6%
Oregon	Fall, 2012	1468	756	51.5%
Washington	Fall, 2012	1462	715	48.9%
Survey Response Bias, California				
	Demographic Variable	Survey Sample	Census Estimates[a]	
	Mean Age (Over 18)	48.0	47.1	
	Median Household Income	$50,000 to $74,999 (Mean category: 5.86)	$60,883 (2006–2010, adjusted average)	
	Gender (Over 18)	Male 49.0%, Female 51.0%	Male 49.5%, Female 51.5%	
	Associates Degree or Higher (Over 18)	49.1%	36.7%	
Survey Response Bias, Oregon				
	Demographic Variable	Survey Sample	Census Estimates[a]	
	Mean Age (Over 18)	49.9	49.5	
	Median Household Income	$50,000 to $74,999 (Mean category: 5.52)	$49,260 (2006–2010, adjusted average)	
	Gender (Over 18)	49.5% Male, 50.5% Female	48.4% Male, 51.6% Female	
	Associates Degree or Higher (Over 18)	40.3%	35.0%	
Survey Response Bias, Washington				
	Demographic Variable	Survey Sample	Census Estimates[a]	
	Mean Age (Over 18)	48.8	48.5	
	Median Household Income	$50,000 to $74,999 (Mean category: 5.70)	$57,224 (2006–2010, adjusted average)	
	Gender (Over 18)	45.9% Male, 54.1% Female	48.7% Male, 51.3% Female	
	Associates Degree or Higher (Over 18)	47.2%	38.8%	
Survey Response Bias, Idaho				
	Demographic Variable	Survey Sample	Census Estimates[a]	
	Mean Age (Over 18)	50.7	48.0	
	Median Household Income	$50,000 to $74,999 (Mean category: 5.19)	$46,890 (2006–2010, adjusted average)	
	Gender (Over 18)	Male 51.0%, Female 49.0%	Male 50%, Female 50%	
	Associates Degree or Higher (Over 18)	40.6%	39.1%	

[a]Data obtained from the U.S. 2010 American Community Survey Public Use Microdata Sample

in the population than in the sample had completed an associate's degree or higher level of education.

Overall, then, the public samples used in this study were generally representative of the known attributes of the populations of the four states, with the exceptions in education levels and income as noted. But, where appropriate, the bias that might be introduced in our overall picture of the specific states or in terms of the relationship of our core variables (AET/NEP) to other variables, will be managed by controlling for income and education.

The core analysis on which the remainder of the book relies is focused on two fundamental concepts. The first is the AET-Tradeoff Inventory (AETTI), which was developed specifically for this study. The second is the New Environmental (ecological) Paradigm (NEP). The NEP is a concept developed several decades ago and which, in the search to facilitate understanding of the environmental/green revolution of the 1960 and 1970s, served to lubricate a major change in the way social scientists looked at individuals' relationship to the environment.

7.3 The Alternative Energy Technology-Environmental Tradeoff

In order to assess the degree to which individual citizens in the different states support the installation of AETs, *even if some local environmental negative consequences occur,* the project developed the series of "trade-off" questions shown in Table 7.2. The particular tradeoff examples were selected from actual AET siting conflicts. These tradeoff scenarios are drawn from stories on the front pages of daily newspapers, online blogs from around the country, and information derived from the case studies in Chaps. 3, 4, 5 and 6, indicating their presence in the minds of the public. Table 7.3 shows the state means on the four trade-off items, as well as on the overall index created by summing (after appropriate recoding) for each individual across the four items. For the overall index, the higher the score (20 maximum) the less likely is the individual to support AETs if they cause environmental harm; conversely, the lower the score (4 minimum), the more likely is the respondent to support AETs even if they damage the environment. Factor analysis of the relationships among the items showed that they reflected various positions on a single dimension. Only one factor emerges with an eigenvalue of greater than 1.0 (1.808) and it explains 45% of the variance in the responses. For the total populations, all four items load greater than .55 and two of them have loadings greater than .72 (wave and hydro items). The factor structures are generally the same in each of the four states when the analysis is conducted separately. The one exception was for the solar tradeoff question, which had a substantially lower loading in Oregon and Idaho (.456 and .510 respectively) than in California (.619) and Washington (.621). Cronbach's scale analysis of an additive index of the four items for the total

Table 7.2 Survey questions

a.	Develop high potential wind energy "farm" in windy rare "tall grass prairie" area.			Preserve rare "tall grass prairie" ecology even if this prevents wind energy farm production.	
	1	2	3 (neutral)	4	5
b.	Leave undisturbed desert floor ecology that supports endangered turtle species, even if it prevents solar power installation.			Install solar power farm in desert even if it disrupts ecology of endangered turtle.	
	1	2	3 (neutral)	4	5
c.	Install off-shore wave turbines on the sea floor to produce electricity even if it impedes whale migration.			Protect whale migratory routes even if it means prohibiting installation of off-shore wave energy turbines.	
	1	2	3 (neutral)	4	5
d.	Continue to operate hydroelectric dams for power even if it harms salmon runs.			Remove dams where they cause harm to salmon runs.	
	1	2	3 (neutral)	4	5

In recent years there has been considerable attention and development of what many call *alternative energy technologies,* such as solar power, wind power, wave and tidal energy and biofuels. Sometimes, though, alternative energy technologies have impacts that some people find undesirable. We are interested in *your views* of what should be done in cases where those impacts are likely to occur. Please circle the number in each case that best represents your opinion. [1 = Strongly Agree; 2 = Agree; 3 = Neutral; 4 = Agree; 5 = Strongly Agree]

Table 7.3 State mean scores on AET tradeoff questions[a]

Topic	California	Oregon	Washington	Idaho	F	Sig.
Wind Farms	2.73	2.80	2.75	2.79	0.76	0.52
Solar Farms	2.79	2.92	2.88	2.81	1.94	0.13
Off-shore Turbines	3.52	3.34	3.50	3.36	4.75	0.003
Hydroelectric	2.91	2.90	2.89	2.85	0.45	0.72
Overall Index	11.95	11.96	12.01	11.82	0.48	0.70

[a]Lowest score (1) is pro-Alternative Energy Technology; highest score (5) is pro-environmental protection. The direction of solar energy trade-off responses has been recoded to be consistent with the items. F-tests and significance were produced by one-way analysis of variance

population showed strong reliability, with an alpha of .592, and it correlates with the factor score result at r = .876. In the following analysis, we employ the index score rather than the factor score because of the more apparent direct connection to the questions themselves. The index kurtosis is .062 and its skewness is .254.

As Table 7.3 shows, the overall index scores were very similar across the four states. Only one item showed a significant difference among the states, namely the one concerning offshore power turbines. Oregon and Idaho respondents were less likely than California and Washington respondents to take the pro-environment

Table 7.4 Correlations (r) of summary tradeoff index with demographic variables, by state and total sample

Demographic variable	California	Oregon	Washington	Idaho	Total index
Age	−.05	−.07	−.10*	−.02	−.06**
Gender	−.14**	−.11**	−.23**	−.15**	−.16**
Education	.03	.07	.04	.01	.04*
Time/Community	−.04	−.06	−.13**	−.03	−.06**
Employment	−.10*	−.01	−.16*	−.05	−.08**
Income	−.08*	−.02	−.07	−.07	−.06**

*p ≤. 05, **p ≤. 01

position, albeit perhaps for different reasons. Offshore power production surely is seen as irrelevant to Idaho, so there likely would be fewer local sources of opposition, nor would there have been any concentrated marketing effort on its behalf. On the other hand, Oregon is the state with the greatest media discussion of offshore turbines, primarily in coastal border counties; thus, the putative negatives surrounding it would be more salient. Even so, in each state, the offshore turbine question had higher average pro-environment scores than the other three AET forms.

The next step in mapping the terrain of the AET tradeoffs was to examine the extent to which support or opposition for the AET position on the index was related to the respondents' demographic attributes. Table 7.4 reports the correlations (r) between the overall Trade-off index and the six demographic attributes of age, gender, education, length of time in one's community, employment status, and income. That table shows the correlations both for the overall sample and for the samples from each of the four states. For the overall sample, each of the demographic variables had a statistically significant correlation (r) with the Tradeoff index, but none of the relationships were particularly strong. The strongest correlation ($r = -.16$, $p = .000$) with the overall index was for the relationship between gender and tradeoffs, with women being more likely than men to take the side of local environmental protection against the perceived harm of the AET. Or, to put it the other way, men were somewhat more likely than women to support the AET even if it causes local environmental harm. That relationship is strongest in Washington State ($r = -.23$**) and weakest ($r = -.11$**) in Oregon.

This analysis also explores the question of whether the effect of individuals' demographic attributes varied among the forms of AET. In some cases, as shown in Table 7.5, the answer was yes. Gender, for example, had a fairly strong correlation with attitudes about protecting the environment from the negative effects of solar, wave, and hydro forms, but had a near-zero correlation with support or opposition to wind power. In fact, across the various demographic attributes, attitudes about wind power AET generally exhibited the lowest correlations. Perhaps this stems from the greater history, presence, and visibility of wind power installations, with accompanying information diffusion significantly greater to the extent that it diminishes the impact of individual variations.

Table 7.5 Correlation of demographic attributes with support for protection against alleged negative environmental effects of alternative energy technologies

	Alternative energy form			
Demographic attribute	Wind	Solar	Wave	Hydroelectric
Age	.05*	−.09**	−.07**	−.07**
Gender	−.02	−.10**	−.14**	−.17**
Education	.05	−.02	.08**	.03
Length of Residence	.03	−.06**	−.06**	−.09**
Employed	.02	−.09**	−.07**	−.08**
Income	−.03	−.09**	−.01	−.01

*p < .05; **p < .01

7.4 New Ecological Paradigm

In the western industrialized democracies, the period after WWII witnessed massive cultural change that transformed politics, cultures, and societies. This period has had various terms applied to it, two of the most frequent and powerful being "postindustrial" and "postmaterial" (or postindustrial and postmaterial) (Bell 1973; Dalton 1988; Inglehart 1997). The explanation for this new political culture grew out of several significant attributes of the post-war period, namely widespread affluence, the growing role of technology as a basis for public policy initiatives, and a general sense of personal security with the easing of the World War II conflict. These attributes enabled western democratic publics to take their physical well-being and personal safety generally for granted, and to focus on what were called "higher order values" (Abramson and Inglehart 1995; Flanagan 1982; Pierce et al. 2000). Among these newly significant "higher" order values for major portions of the populations were the right to participate in decisions affecting them, a growing emphasis on equality for previously disadvantaged segments of the population, a growing opposition to foreign military involvements by major powers, and a new paradigm for viewing natural resources and the environment (Aaron et al. 1994; Dalton 1988; Inglehart and Baker 1999; Pierce et al. 1992).

The term "New Environmental Paradigm" (later re-labeled as the New Ecological Paradigm) was coined by Catton and Dunlap, two sociologists who also were given credit for the creation of a new discipline called "Environmental Sociology" (Dunlap and Catton 1979; Dunlap et al. 2000). The New Environmental Paradigm (NEP) contained a very different view of the relationship between humans and nature than had the "Dominant Social Paradigm" (DSP). With some minor exceptions, this "new" view saw humans and the environment as equal participants in the larger ecological systems of the world (Dunlap 2002). This paradigm was very unlike the DSP, which saw nature as a resource to be exploited for the benefit of humans, one that contained no inherent value or intrinsic rights of its own. This "new" NEP way of looking at the environment and humans emerged hand-in-hand with the rise of widespread environmental movements in many countries, most notably the Western democracies. Coupled with a heightened emphasis on the protection of the environment was also an expression of strong demands for mass publics to have the opportunity to participate in the actual policy decisions affecting natural resources and the environment, such as priorities for the use of rivers, or the efficacy and desirability of clearcutting coniferous forests (Pierce et al. 2000).

Moreover, environmental interest groups experienced mushrooming growth during this period, both enhancing the influence and power of the NEP value orientation and assisting in the embedding of environmental content into school curricula from pre-school to post-graduate work (Pierce et al. 1992; Steel and Warner 2007). Perhaps no changes were more significant in this regard than the formation of the U.S. Environmental Protection Agency (EPA) at the federal level, and the enactment of the National Environmental Policy Act (NEPA) requirement for environmental impact statements to be part of a relevant activity under federal jurisdiction (note the preceding case studies chapters). The present question is whether the attitudes about the tradeoffs involved in AETs derive from fundamental orientations to the environment, and when strongly held, whether those orientations lead members of the public to oppose the AET technologies that also are rationalized in terms of environmental goals, when they also are said to impose their own negative environmental consequences or tradeoffs.

The NEP measure we employed here is based in survey participants' responses to six items drawn from a revised NEP index (Dunlap et al. 2000). Each item asks respondents to give a reaction on a five-point scale, ranging from 1 = strongly disagree to 5 = strongly agree. Since the development of the original NEP measure, the index has been both used and referenced in hundreds of subsequent scholarly publications, including those of which Dunlap has been a part, but also by numerous other researchers (e.g., Hawcroft and Milfont 2010). The NEP may be the most cited and employed measure in the social science study of the environment, and one of the most in all of social science in the past three decades (Lundmark 2007). Nonetheless, the measure has received both criticism from other scholars, as well as modification by Dunlap himself (Amburgey and Thoman 2011; Dunlap et al. 2000). Such critiques usually argue that the NEP measure represents more than a single dimension—but there is no agreement on just how many dimensions are embedded in the measure, nor on what the implications for substantive conclusions are produced by that critique (Cordano et al. 2010). There are different versions of how many questions must be asked of respondents in order to fully represent their views on the relationships of humans and the environment (Amburgey and Thoman 2011). And other scholars wonder whether the NEP is culture specific, and thus not particularly relevant to non-Western countries with cultures, politics, and philosophies in regard to the environment that are very different from those in the United States, where most of the pre-testing of the measure occurred (e.g., Erdogan 2009).

As indicated above, this present study uses a subset of the items in Dunlap's measure of the NEP, relying on 6 questions out of the 15 from the revised version (Dunlap et al. 2000). While there have been some academics concerned about using subsets of the NEP (e.g., not as robust measure as the complete scale; see Hawcroft and Milfont 2010), this study nonetheless moves forward with the six-item NEP for several reasons. First, by far the weight of the research in environmental attitudes

Table 7.6 State mean scores on NEP items and scale

	California	Oregon	Washington	Idaho	F	Sig.
Balance of Nature	3.91	3.85	3.88	3.82	.74	.53
Right to Modify	2.75	2.69	2.70	2.81	1.63	.18
Limit of People	3.49	3.52	3.43	3.46	.53	.66
Ecological Crisis	2.57	2.58	2.58	2.82	5.41	.001
Plants/Animals	3.68	3.72	3.61	3.54	2.91	.033
Rule Over Nature	2.55	2.52	2.58	2.73	3.34	.018
Overall Index	21.21	21.30	21.07	20.46	3.58	.013

N = 2766–2843

and values demonstrates that the NEP has both internal reliability and external validity. Variations on NEP measures are widely seen to be consistent with those predicted by theoretical expectations, including relationships to postmaterial values, attitudes on specific issues of environmental concern, knowledge about environmental issues and environment-related science, and higher levels of involvement in environmental politics (e.g., Steel 1996; Pierce et al. 1992). Second, the authors of this present work themselves have a long history with the NEP, not only in enhanced understanding of environmental issues and policies in the United States, but also in a number of other countries, including Canada, Japan, and the former Soviet Union (e.g., Pierce et al. 1992, 2000). Thus, the respondents in this project's surveys were asked to indicate their level of agreement or disagreement (strongly agree, agree, neutral, disagree, strongly disagree) with the following six statements:

(i) The balance of nature is very delicate and easily upset by human activities.
(ii) Humans have the right to modify the natural environment to suit their needs.
(iii) We are approaching the limit of people the earth can support.
(iv) The so-called "ecological crisis" facing humankind has been greatly exaggerated.
(v) Plants and animals have as much right as humans to exist.
(vi) Humans were meant to rule over the rest of nature.

After reverse recoding of items ii, iv, and vi, the responses also were summed across all six items to produce an overall index, possible scores on which ranged from 5 to 30. Table 7.6 shows the average response for the respondents from each state on each item (before reverse re-coding three items) and for the overall index.

As a test of the integrity of this "abridged" NEP measure, a factor analysis was employed to estimate the degree to which the six questions reflect the same dimension, namely one of variation on the NEP. Results for the four-state aggregated sample produced a single factor, rather than the multiple factors found in some other versions of the measure. In this sample, none of the individual item factor loadings were less than .62. The highest loadings were obtained for the "ecological crisis" (.728) and "rule over nature" (.711) items. Only one factor emerged with an eigenvalue of greater than 1.0 (2.728), an often-used cutoff value, and that factor explains

Table 7.7 Correlations (r) of NEP and tradeoff items

	Tradeoff items			
NEP Items	Wind	Solar	Wave	Hydroelectric
Balance of nature	.13**	.23**	.24**	.34**
Humans can modify	−.18**	−.25**	−.19**	−.29**
Earth at its limit	.09**	.06**	.14**	.22**
Ecological crisis over-rated	−.14**	−.14**	−.25**	−.26**
Plants and animals	.15**	.23**	.18**	.27**
Humans meant to rule	−.17**	−.22**	−.21**	−.25**

**$p \leq .01$; N = 2766–2843

45% of the variance in responses to the items. Thus, the analysis moves forward with the NEP as a core measure of the research.

The overall pattern in Table 7.6 shows that Idaho residents were slightly less supportive of the NEP than were the residents of the other three states. Idaho had the lowest average score on the overall summative NEP index. Moreover, Idahoans were somewhat more likely than those in the other three states to believe that "the so-called 'ecological crisis' facing humankind has been greatly exaggerated," somewhat less likely to believe that "plants and animals have as much right as humans to exist," and more likely to believe that "humans were meant to rule over the rest of nature."

One question of interest is whether specific NEP components are more central to the formation of attitudes about alternative AETs than are others. Thus, Table 7.7 shows the correlations among the individual items in the two sets—AET and NEP. The strongest correlation in the set shown in Table 7.7 was between agreement that "The balance of nature is very delicate and easily upset by human activities" and the AET position to "Remove dams where they cause harm to salmon runs." There undoubtedly is significant logic to this pattern, given the long history of conflict in each of the four states over the obvious negative effect of hydroelectric facilities on *anadromous* fisheries (those fish, such as salmon, that spawn in upstream fresh water, the smolt migrating downstream to mature in salt water before returning to fresh water to spawn and die). Those fish must pass through the dams going in both directions– downstream as smolt heading to the sea, and upstream heading to spawn. This impact has led to the removal of several dams on some rivers and streams. And, indeed, the hydroelectric AET question has the strongest relationship overall to each of the NEP items.

Moreover, as shown in Table 7.8, when controlling for the state of the respondent, the same pattern holds course. That is, in all four states the strongest correlations between the individual AET tradeoff questions and the NEP index score were for hydroelectric power. At the same time, for solar power the strongest correlation was in California, where the region's greatest installation of solar facilities are found. Idaho and Oregon publics' attitudes about wind power installation were more closely tied to their NEP support than was found for Washington and California. In the case of Oregon, once again, this may stem from the salience of wind power and the resulting controversy over its installation. That is, in 2010, 7.1%

Table 7.8 Correlations of NEP index and AET trade-off items by state

	California	Oregon	Washington	Idaho	Overall
Wind	.16**	.26**	.17**	.27**	.21**
Solar	.34**	.23**	.28**	.26**	.28**
Wave	.29**	.36**	.22**	.32**	.30**
Hydroelectric	.43**	.44**	.34**	.40**	.40**
Tradeoff index	.45	.48**	.38**	.46**	.44*

N = 2732
*p < .05; **p < .01

of power in the state of Oregon was produced by wind, compared to 3%, 3.7% and 4.6% respectively for California, Idaho, and Washington. The numerous controversies over Oregon's development of wind power no doubt have served to crystallize those attitudes and clarify their implications for the environment.

Overall then, in a statistical context, individuals' positions on the NEP were significantly related to their positions on the trade-offs involved in the development of AETs and those relationships vary some among the states in a way consistent with their AET politics. At the same time, though, there appears to be enough unexplained variation in the pair-wise relationships to suggest that the answer is more complex than that. Indeed, why do some environmentalists support alternative energy technologies when others do not?

One major focus of this book is the attempt to understand the prevalence of various combinations of support or opposition to the installation of AETs and high and low support for the NEP. Thus, in these publics, how many respondents express strong environmental values on the NEP, yet also oppose AETs designed to reduce impact on the environment when those technologies are said to have their own negative environmental impact? Likewise, how many low NEP respondents are nonetheless moved to oppose the AETs with negative environmental impacts of their own? To facilitate that inquiry, both measures were partitioned into relatively equal thirds, based on the distribution of respondents. The cutting points for the AET Tradeoff index were as follows: low =10 and below (32%); medium =11 and 12 (31%) and high =13 and greater (38%). For the NEP index, the cutoff points were as follows: low =18 and below (32%); medium =19 through 24 (38%); and high =25 and higher (30%). Thus, one is able to compare individuals who are HighAET/LowNEP to those who are HighNEP/HighAET, or LowAET/LowNEP or LowAET/HighNEP, as well as those who fall somewhere in between those extremes.

The first look at these combinations is found in Table 7.9, which shows that the most frequently populated cells of the combinations of the AET and the NEP are found in the upper left and lower right corners of the table. Those respondents who might appear to be consistent in their AET and NEP orientations are in the cells with low NEP/low AET (6%) or high NEP/high AET (4%) combinations. At the same time, though, it is the group in that lower right corner (with the largest cell percentage) that seems to be making the most paradoxical choice—opposing the pro-global-environmental AET technologies because of the more local, site-specific anticipated negative environmental effects. On the other hand, 10% of the

Table 7.9 Trade-off/NEP types

AET/Tradeoff	NEP Types Low	Medium	High
HighAET	16%	12%	4%
MedAET	10%	12%	9%
LowAET	6%	15%	17%

N = 2732

respondents are in the other corners of the table—those 4% high on the NEP measure, while also high on the support of alternative energy technologies, even if they have negative site-specific environmental consequences, or those 6% with very low support for the NEP, who also oppose alternative energy technologies when they have negative environmental consequences.

State comparisons of the percentage of the various combinations of AET and NEP revealed some small differences. Idaho had a slightly higher percentage (18%) of those who were low environmental oriented on both the AET and the NEP than did Oregon (14%), California (15%) and Washington (16%). Consistent with that, Idaho also had the lowest percentage that is high on both measures—pro-AET *and* high NEP). Idaho also had a slightly higher percentage that opposed the alternative energy projects, even while they were in the lowest category of the NEP, suggesting perhaps some other reasons for that opposition. Oregon, on the other hand, has the greatest percentage that is very high on the NEP, but also very supportive of the AETs.

Small sub-groups in the state samples held seemingly contradictory general orientations to the issues produced by the siting of alternative energy technologies—e.g., those with high NEP values who also were supportive of alternative energy technologies that may pose a threat to certain elements in the environment (who might be called global-environmentalists or "globalists"), or those who support AETs designed to deal with carbon-induced climate change, but who were at the low end of the NEP value index. The question may arise as to whether these combinations differ in other ways that may affect policy outcomes. Are people from these groups more likely to feel as though they can make a difference and therefore presumably more likely to participate in discussions and policy formation activities? As will be shown in Chap. 8, respondents (whom we have call "localists") who were both high on the NEP scale and opposed the alternative energy sites when they had a putative negative environmental impact were the most likely to talk about renewable energy policy and they were more likely to believe that people like them are able to have an impact on state energy policy.

On the other hand, while respondents who were low on the NEP index and higher in support for alternative energy installations even with negative environmental impacts (the rationalists), were much less likely than the other three groups to feel that people like them can have an impact on state energy policy. The two mixed-extreme groups were less likely than either of the two consistent extreme groups to talk about renewable energy issues, but they were more likely than the low-NEP/pro-AET group to feel that people like them can have an impact on state energy

Table 7.10 Regression of AET scales on NEP and demographic variables

	Summary index	Wind	Solar	Wave	Hydro
	b	*b*	*b*	*b*	*b*
Age	.07***	.09***	.02	.02	.08***
Gender	.07***	.02	−.02	.10***	.10***
Education	.02	.06***	−.04*	.06**	−.01
Residence	−.04	.02	.04	.00	.09***
Employment	.09***	−.02	.11***	−.04	−.07**
Income	−.02	−.01	.08***	.02	.03
NEP	.43***	.24***	.26***	.29***	.38***
F=	96.0***	23.3***	36.9***	45.4***	76.7***
R^2=	.21	.06	.09	.11	.18
N=	2527	2531	2544	2540	2539

policy. And, the High NEP/Pro-AET respondents (the globalists) were less likely than the High NEP/Low AET respondents (localists) to both talk about renewable energy and to feel that they have an impact on state energy policy. This sense of efficacy, the feeling that one can have an impact on policy outcomes, may be a key to the impact of the environmentalist opponents of specific AETs.

To this point, the analysis in this chapter has shown statistically significant relationships between the New Ecological Paradigm and opposition to alternative energy technologies when they are said to have negative environmental consequences. The results also have shown that there are numerous statistically significant, albeit small effect size, correlations between demographic characteristics of respondents and positions on the alternative energy technology questions, and that the "localists" (pro-NEP/Anti-AET) were more likely to talk about renewable energy and to feel they have an impact on state energy policy. Given those patterns, the analysis now turns to whether the NEP retains its effect on positions on AETs when controlling for those demographic attributes.

Table 7.10 regresses the AET scale on six demographic variables and the NEP. While a number of the intersections of the demographics and the AET measures showed significant relationships, overall the most powerful effect was found with the NEP. Thus, even when controlling for the demographics, the taking of a position on the AETs appeared to be largely determined by one's overall environmental/ecological values. That is, individuals with strong environmental values were more likely to oppose AETs if the installation of those technologies is thought to cause environmental damage of another, and perhaps more specific kind. But, there also remains significant unexplained variation in AET opinions that is not accounted for by the NEP, namely the *globalists* who express pro-environmental values (high NEP), but seem to see those concerns globally as more significant than the local environmental effects. This is the question that the remainder of the book will address: Why is it that individuals who *share* a commitment to environmental values may *differ* in terms of how those values are applied in the cases of the installation of alternative (renewable) energy technologies?

7.5 Conclusion

The central focus of this chapter has involved an examination of the role of environmental values—most specifically, the New Environmental Paradigm—in the shaping of citizen orientations toward environmental impacts versus renewable energy technology tradeoffs. More specifically, the New Ecological Paradigm index (in its six-item form) was used as a way of representing value-based orientations to relationships between humans and the natural world (Dunlap et al. 2000). Findings here show that certain shared environmental values are indeed held by individuals who either support or oppose the development of AETs versus local environmental conditions. To be sure, those respondents with higher NEP scores were more supportive of local environmental AET considerations than were individuals with lower NEP scores, but there also were substantial pro-NEP respondents who were more prone to supporting AET development, even if there were site-specific negative environmental consequences. The Pro-NEP/Anti-AET group also was more likely to talk about renewable energy policy and was more likely believe they can have an impact on state energy policy.

Taken together, these three patterns (pro-NEP, more talk, and more efficacy) may provide insight into the reasons the installation of AETs such as wind farms or solar farms often face very effective opposition, as was shown in the earlier case studies case studies. The following chapters will explore these patterns in greater detail. So, the question remains, what is there about the AETs that leads some environmentalists to oppose the AET siting? The following chapters explore that question in regard to such possible explanations as their more general values, their particular political orientations, their specific beliefs about energy policy, their knowledge levels, where they obtain their policy-relevant information, and their real or hypothetical proximity to AET installations.

References

Aaron, H., Mann, T. E., & Taylor, T. (Eds.). (1994). *Values and public policy*. Washington, DC: The Brookings Institution.

Abramson, P., & Inglehart, R. (1995). *Value change in global perspective*. Ann Arbor: University of Michigan Press.

Amburgey, J. W., & Thoman, D. B. (2011). Dimensionability of the new ecological paradigm: Issues of factor structure and measurement. *Environment and Behavior, 44*, 235–256.

Bell, D. (1973). *The coming of post-industrial society*. New York: Basic Books.

Bell, D., Gray, T., & Haggett, C. (2005). The social gap in wind farm siting decisions: Explanations and policy responses. *Environmental Politics, 14*, 460–477.

Bidwell, D. (2013). The role of values in public beliefs and attitudes towards commercial wind energy. *Energy Policy, 58*, 189–199.

Cordano, M., Welcomer, S. A., & Scherer, R. F. (2010). An analysis of the predictive validity of the new ecological paradigm scale. *The Journal of Environmental Education, 34*, 22–28.

Dalton, R. (1988). *Citizen politics in western democracies: Public opinion and political parties in the United States, Great Britain, West Germany and France*. Chatham: Chatham House.

Dillman, D. A. (2010). *Internet and mail surveys: The tailored design method* (2010th ed.). New York: Wiley.
Dunlap, R. E. (2002). An enduring concern. *Public Perspective, 13*, 10–14.
Dunlap, R. E., & Catton, W. R. J. (1979). Environmental sociology. *Annual Review of Sociology, 5*, 243–273.
Dunlap, R. E., Van Liere, K., Mertig, A., & Jones, R. (2000). Measuring endorsement of the new environmental paradigm: A revised NEP scale. *Journal of Social Issues, 56*, 425–442.
Ellis, G., Barry, J., & Robinson, C. (2007). Many ways to say "no", different ways to say "yes": Applying Q-Methodology to understand public acceptance of wind farm proposals. *Journal of Environmental Planning and Management, 50*(4), 517–551. doi:10.1080/09640560701402075.
Erdogan, N. (2009). Testing the new ecological paradigm: Turkish case. *African Journal of Agricultural Research, 4*, 1023–1031.
Flanagan, S. (1982). Changing values in postindustrial society. *Comparative Political Studies, 14*, 99–128.
Gross, C. (2007). Community perspectives of wind energy in Australia: The application of a justice and community fairness framework to increase social acceptance. *Energy Policy, 35*(5), 2727–2736. doi:10.1016/j.enpol.2006.12.013.
Grover, S. (2012, January 19). Tradeoff between environment and economy is "Nonsense" says GE executive. *Treehugger*. http://www.treehugger.com/corporate-responsibility/tradeoff-between-environment-and-economy-nonsense-says-ge-executive.html. Accessed 15 Aug 2016.
Haggett, C. (2011). Understanding public responses to offshore wind power. *Energy Policy, 39*, 503–510.
Hawcroft, L., & Milfont, T. (2010). The use and abuse of the new environmental paradigm scale over the last 30 years: A meta-analysis. *Journal of Environmental Psychology, 30*, 143–158.
Inglehart, R. (1997). *Modernization and postmodernization: Cultural, economic, and political change in 43 societies*. Princeton: Princeton University Press.
Inglehart, R., & Baker, W. (1999). Modernization, cultural change, and the persistence of traditional values. *American Sociological Review, 65*, 19–51.
Kraft, M. E., & Clary, B. B. (1991). Citizen participation and the Nimby syndrome: Public response to radioactive waste disposal. *Political Research Quarterly, 44*(2), 299–328. doi:10.1177/106591299104400204.
Leiserowitz, A., Maybach, E., Roser-Renouf, N., & Smith, N. (2011). *Global Warming's Six Americas*. New Haven: Yale Project on Climate Change. https://environment.yale.edu/climate-communication/files/SixAmericasMay2011.pdf. Accessed 23 Dec 2015.
Lundmark, C. (2007). The new ecological paradigm revisited: Anchoring the NEP scale in environmental ethics. *Environmental Education Research, 13*(3), 329–347.
Pasqualetti, M. J. (2011). Opposing wind energy landscapes: A search for common cause. *Annals of the Association of American Geographers, 101*(4), 907–917. doi:10.1080/00045608.2011.568879.
Petrova, M. A. (2013). NIMBYism revisited: Public acceptance of wind energy in the United States. *Wiley Interdisciplinary Reviews: Climate Change, 4*(6), 575–601. doi:10.1002/wcc.250.
Phadke, R. (2010). Steel forests or smoke stacks: The politics of visualisation in the Cape Wind controversy. *Environmental Politics, 19*(1), 1–20.
Pierce, J. C., Steger, M. A., Steel, B. S., & Lovrich, N. P. (1992). *Citizens, political communication, and interest groups: Environmental organizations in Canada and the United States*. New York: Praeger Publishers.
Pierce, J. C., Lovrich, N. P., Steel, B. S., & Steger, M. A. (2000). *Political culture and public policy in Canada and the United States: Only a border apart?* New York: Edwin Mellen Publishers.
Pierpont, N. (2009). *Wind Turbine syndrome: A report on a natural experiment*. Lowell: King Printing.

Sebelius, S. (2013, August 13). Green energy is good, but it comes with a trade-off. *Las Vegas Review-Journal*. http://www.reviewjournal.com/columns-blogs/steve-sebelius/green-energy-good-it-comes-trade. Accessed 15 Aug 2016.

Steel, B. S. (1996). Thinking globally and acting locally?: Environmental attitudes, behavior and activism. *Journal of Environmental Management, 47*, 27–37.

Steel, B. S., & Warner, R. L. (2007). Global academia: The state of environmental learning and awareness. In K. V. Thai, D. Rahm, & J. D. Coggburn (Eds.), *Handbook of Globalization and the Environment* (pp. 233–255). Philadelphia: Taylor and Francis.

Warren, C. R., Lumsden, C., O'Dowd, S., & Birnie, R. V. (2005). "Green on green": Public perceptions of wind power in Scotland and Ireland. *Journal of environmental planning and management, 48*(6), 853–875.

Wolsink, M. (2007). Wind power implementation: The nature of public attitudes: Equity and fairness instead of "backyard motives.". *Renewable and Sustainable Energy Reviews, 11*(6), 1188–1207.

Wolsink, M. (2010). Near-shore wind power—Protected seascapes, environmentalists' attitudes, and the technocratic planning perspective. *Land Use Policy, 27*(2), 195–203.

Chapter 8
A Matter of Values?

Abstract This chapter examines more closely the impact of individuals' postmaterial value orientations on public support for or opposition to development of renewable energy resources such as biofuels, wind, geothermal, solar and wave energy. More specifically, the following analysis examines how postmaterialist values affect perceptions of potential environmental tradeoffs between the benefits and potential costs of development and siting renewable energy technologies. The central question is whether the development of postmaterial society and subsequent higher order needs (e.g., those resulting in biocentric orientations toward the environment) are related to support for, or opposition to, the development of alternative renewable energy technologies. Using both the postmaterialist value indicator developed by Ronald Inglehart and his colleagues (Inglehart 1991, 1997) and the New Ecological Paradigm index developed by Dunlap and his colleagues (Dunlap et al. 2000), this chapter examines the impact of values on preferred environmental tradeoffs concerning different renewable energy technologies. Finally, the chapter assesses more directly whether the localists and globalists differ from each other on those values.

Keywords NIMBY • New Ecological Paradigm • Postindustrial society • Postmaterialist values

8.1 Introduction

> *If we cannot put solar power plants in the Mojave Desert, I don't know where the hell we can put it* (Governor Arnold Schwarzenegger)
>
> *You couldn't put a project in a worse area from a landscape connectivity point of view… It's a philosophic non-sequitur that you can destroy hundreds of thousands of acres to save the Earth from global warming* (David Meyers, Director, Wildland Conservancy (Woody 2010))

The lack of social acceptance of renewable energy technology and siting by affected communities is considered one of the most significant barriers to their

J.C. Pierce, B.S. Steel, *Prospects for Alternative Energy Development in the U.S. West*, Environmental Challenges and Solutions 4,
DOI 10.1007/978-3-319-53414-5_8

development (Petrova 2013; Strachan and Lal 2004). Thus, to facilitate that siting is important to understand what motivates individuals and communities to support or oppose that development. The vast majority of research in this area suggests that NIMBYism (Not in my Back Yard), while a frequently given reason, is not a sufficient explanation for understanding public opposition to the development of utility-scale alternative energy technologies (AETs), such as wind or solar farms (Petrova 2013). Moreover, as the previous chapter demonstrated, given the number of environmental value holders who oppose AET sitings, the holding of environmental values is perhaps a necessary, but not sufficient explanation for support for those technologies. Prior work has identified visual and landscape concerns (e.g., Devine-Wright 2010, 2011; Haggett 2011; Pasqualetti 2000, 2011), socioeconomic concerns (Wolsink 2010), procedural concerns (e.g., Devine-Wright 2010, 2011; Haggett 2011, 2012), and environmental value concerns (Slattery et al. 2012; Steel et al. 2015). It is the role of values with which this chapter is most concerned.

As discussed earlier, one would expect to find greater support for AETs among those who hold biocentric and postmaterialist values, and greater opposition to those technologies among those who are more anthropocentric, with more traditional or materialistic values. Chap. 7 examined the link of support for AETs with the New Environmental Paradigm values. Likewise, people who believe the now abundant scientific evidence on global climate change would be expected to be more likely to favor the development of renewable energy sources, while those who are suspicious of this evidence are likely to view social investments as wasteful and a reflection of misguided public policy (McCright and Dunlap 2011). Such bundles of shared values seem likely to be associated with either support for or opposition to the promotion and implementation of AETs. Indeed, recent research reports ample evidence of non-science based, values-inspired cultural framing of the increasingly convincing evidence on anthropogenic global climate change. That values framing can make the social marketing of governmental programs a challenging undertaking in many countries, including the U.S. (Corner and Randall 2011; Fischer et al. 2011).

As the prior chapters have noted, the more interesting and less expected sources of opposition to AETs are people who fear adverse consequences to local ecological systems in which renewable energy projects are ultimately sited (Devine-Wright 2010), many of whom are environmentalists (Robinson 2004). Indeed, environmentalists who reside in or near areas where renewable energy projects are sited frequently identify harmful side effects of alternative renewable technologies such as wind power, hydropower, geothermal, and wave energy. Thus, in such cases, wind farms sometimes are viewed as a danger to bats and raptors (Johnson et al. 2003; Kikuchi 2008), wave energy farms are thought to pose potential impediments to migrating fish and whales (Cada et al. 2007), and geothermal projects are believed by some to increase the risk of seismic activity (Ferris 2012).

8.2 Postindustrialism and Its Relevance to Renewable Energy

As an affluent postindustrial nation, the United States also depends on a very high rate of use of natural resources—rates unprecedented in human history. This form of economic development has led to a high level of material consumption by a large proportion of the population, a population with a relatively high standard of living that is intimately tied to natural resource exploitation and energy consumption that results in significant adverse impacts on air, water, and land-use practices. As the energy consumption data for the four western states (Chap. 2 and in Appendix A) indicate, the overall level of energy consumed and the pollutants produced are staggering by any measure. Thus, as Gary Bryner has commented, such levels of consumption create an imperative and obligation by all states and nations to take a leading role in global environmental protection efforts (Bryner 1997). In regard to the United States, Bryner argues, "The United States is so economically and politically powerful that its participation in global environmental protection efforts is essential" (Bryner 1997, p. 9).

It is important to note, however, that it is that same high level of technological development—which has led to the high consumption economy and high standard of living—that also has produced sophisticated approaches to the task of cleaning up the environment, addressing natural resource problems, and the development of renewable energy technologies. For example, high-tech particle precipitators are available to be installed on industrial and energy plant smokestacks to prevent many types of air pollution. Similarly, renewable energy technology is leading toward the production of ever-more-efficient wind turbines and solar technologies (Simon 2007).

As a postindustrial nation, the United States also enjoys high levels of economic affluence and technological development while manifesting sustained political stability. One characteristic that all postindustrial nations have in common is that economic development has evolved through each of the various phases of modernization. The economy was first characterized by highly labor-intensive agriculture and rural lifestyles, then by high output of manufactured goods, increasing urbanization, and strong economic growth; finally, the economy was characterized by the predominance of service and tertiary sector activity (see Table 8.1). A decisive feature of this

Table 8.1 Structure of GDP for case study states

	Agriculture/forestry/fishing/mining		Industry/manufacturing		Services	
	% 1980	% 2010	% 1980	% 2010	% 1980	% 2010
California	5.3	3.2	36.4	26.2	58.3	70.6
Idaho	3.7	2.2	34.6	27.2	61.6	70.6
Oregon	4.2	2.3	33.7	26.2	62.0	71.5
Washington	5.8	2.6	39.3	30.5	54.9	66.9
U.S.	3.7	1.7	41.9	37.2	54.4	61.1

Source: U.S. Census Bureau

development was that, for the first time in history, large portions of the general population was able to partake in the substantial material benefits of this economic expansion, namely economic security and well-being.

Because of unprecedented and rapid economic growth in the post-World War II period, due to the increased importance of modern technology, socioeconomic and political structures of these societies were altered, and social commentators began to take note of a new, unprecedented stage of development (Dalton 1988). This new stage or phase of socioeconomic development is commonly referred to as "postindustrial" or "advanced industrial" in the social science literature on modernity and postmodern intellectual thought (Inglehart 1997).

Many studies have examined the social, psychological, economic, and political implications of postindustrialism (Huntington 1974; Inglehart 1997; Inglehart and Baker 1999).). A central theme of postindustrial change identified by these studies is the crucial role of technology. As Ronald Inglehart stated (Inglehart 1997, p. 8), "Technology is creating the Post-industrial society just as it created industrial society." Pierce et al. have argued that technology has led to the following characteristics of this new type of society (Pierce et al. 1992, p. 9):

> It has created new kinds of knowledge as sources of power and influence; it has become the core of both national and international communication processes, revolutionizing the speed and quality of information transmission; technology has become an integral part of the postindustrial culture and provides independence from many traditional political and social institutions; technology has created new occupational structures and the need for expertise to fill those positions in that structure; and, technology has created new centers of power and new resources for power brokers.

Russell Dalton has further suggested that increased educational opportunities for citizens in postindustrial societies have led to "a growth in political skills and resources, producing the most sophisticated publics in the history of democracies" (Dalton 1988, pp. 5–6).

Observers of this postindustrial change have also noted that the nearly universal satisfaction of basic subsistence needs such as food and shelter has altered individual value structures among many citizens, particularly younger cohorts in Western democracies. "Higher order needs" are said to have supplanted more fundamental subsistence needs (e.g., security, economic access) as the motivation for much societal behavior (Flanagan 1982; Inglehart 1991; Abramson and Inglehart 1995; Yankelovich 1989). Value changes entailing greater attention to "postmaterialist," "post-scarcity," or "diminishing material utility needs" are seen to have brought about concurrent shifts in many types of personal behavior, including those related to the environment (Inglehart 1997). Indeed, many have argued that the values associated with modern environmentalism were made possible by this postmaterial value change (Ronald Inglehart 1995; Pierce et al. 2000).

Postindustrial societies have also led to the increased importance of urban or metropolitan areas, with a concomitant decline occurring in rural areas. This shift in population growth and economic activity has serious implications for environmental policy, such that urban areas have gained more political and economic leverage at the expense of rural areas. Finally, with the advent of the greater presence of postmaterialist value orientations among the citizens residing in urban areas, greater numbers of urban mass publics and elites have come to hold belief systems concerning environmental policy that are profoundly at odds with long-held views often found in the more traditional rural settings. As was seen in the case studies, some of this dynamic was at work in siting controversies where, for example, urban constituencies opposed renewable energy projects that potentially would have endangered habitat and species (e.g., Calico Solar, Chap. 3), while supporting such projects when only rural residents' aesthetics and property value concerns were evident (e.g., Kittitas Valley Wind, Chap. 4).

8.3 Postindustrial Value Change

Again, many scholars have argued that that the advent of postindustrial society has led to the alteration of individual value structures among citizens born and raised in times of peace and affluence. These citizens, it is said, have developed "higher order" needs, including a concern for the quality of life, which have supplanted concerns about many basic needs (Flanagan 1982; Aaron et al. 1994; Inglehart 1991, 1997; Inglehart and Baker 1999; Yankelovich 1989). These same observers of postindustrial societies also suggest that the development of the environmental movement was, in considerable measure, a predictable product of the vast socioeconomic changes evident in postwar postindustrial societies (Caldwell 1992; Inglehart 1995; Milbrath 1984). The development of a strong environmental consciousness and the advent of environmental movements in postindustrial societies have resulted in the open questioning of many of the established, longstanding political and economic institutions of modern society (Habermas 1981; Offe 1984). These changing value orientations among individuals and social groups, including some elements of the elite in postindustrial societies, challenge existing environmental policies and land use patterns and have led in time to the articulation of the New Ecological Paradigm (Dunlap et al. 2000).

The Gallup Poll data presented in Table 8.2 indicate that, since 2006, the plurality of Americans in most years prefered "protection of the environment" over "energy development" when given a choice between these two priorities of public policy. These survey data make a case for the existence of widespread public concern for the environment and for likely broader support of the New Environmental Paradigm in postindustrial societies.

While these survey data indicate that much of the public is concerned about environmental issues and wants to protect the environment, what can be said of their actual behavior? The Gallup Poll survey data presented in Table 8.3 provides some indication that many citizens are indeed pursuing environmentally oriented behaviors that reflect that environmental consciousness. For example, in the United States 90% of respondents in 2010 indicated that they are recycling household items. In addition, 85% of the Americans surveyed claim to have reduced their energy usage, 81% say they have replaced standard light bulbs with compact fluorescent light

Table 8.2 U.S. public opinion: energy development versus the environment

Year	% Protection of the environment	% Development of U.S. energy supplies	% No opinion
2015	49	39	12
2014	51	40	9
2013	46	45	9
1912	44	47	9
2011	41	50	9
2010	55	39	6
2009	47	46	7
2008	50	41	9
2007	58	34	8
2006	49	42	9

Source: Gallup Poll (2015)

With which one of these statements about the environment and energy production do you most agree: (1) protection of the environment should be given priority, even at the risk of limiting the amount of energy supplies—such as oil, gas and coal—which the U.S. produces; or, (2) Development of U.S. energy supplies—such as oil, gas and coal—should be given priority, even if the environment suffers to some extent

Table 8.3 U.S. public self-reported environmental behaviors

	2000	2003	2007	2010
Self-reported behaviors	% yes	% yes	% yes	% yes
Voluntarily recycled newspapers, glass, aluminum, motor oil, or other items	90%	89%	89%	90%
Reduced your household's use of energy	83%	80%	85%	85%
Replaced standard light bulbs in your home with compact fluorescent light bulbs				81%
Bought product specifically because you thought it was better for the environment	73%	72%	70%	76%
Used reusable shopping bags at grocery stores instead of the standard plastic or paper bags				70%

Source: Gallup Poll (2010)

bulbs, 76% say they practice "green consumerism" by buying environmentally responsible products and 70% say they used reusable shopping bags instead of plastic or paper bags.

These survey data indicate the presence of many values and behaviors concerning protection of the environment and may suggest general support for the New Ecological Paradigm. However, it remains the case that a culture of consumerism continues to dominate the lives of most citizens living in contemporary postindustrial societies. Carl Boggs describes life in postindustrial America as "an unprecedented resort to personal consumption, which finds ready expression through the countless images circulated in the spheres of advertising, mass media, popular culture, sports and fashion" (Boggs 2000, p. 80). According to this view, postindustrial nations are well on their way to ecological catastrophe through an insatiable desire for material items such as electronics, cosmetics, videos, and multiple cars. To fulfill all of these material wants, large amounts of natural resources and energy consumption are required, and the associated production processes often pour pollutants into the environment.

The work of Ansolabehere and Konisky is most informative here. In *Cheap and Clean: How Americans Think About Energy in the Age of Global Warming*, they find from a series of surveys conducted over a decade (Ansolabehere and Konisky 2014, pp. 18–19): "Ultimately, people want lower prices, a cleaner environment, and more reliable energy. Those preferences are strong. They are always present. They drive the energy sector to continually innovate. And they are ultimately the source of hope that we as a nation and a world can address the problems of global climate change embedded in our use of energy."

8.4 Measuring Postmaterial Values

The most important measure of postmaterialist values was developed in the work of Ronald Inglehart (1991). This measure is used in the analysis below. Each survey respondent was asked to choose two statements of four as their first and second most important goals for "our country" for "the next ten or fifteen years." The four alternative goals were as follows: (a) Maintaining order in the nation; (b) Giving people more say in important governmental decisions; (c) Fighting rising prices. (d) Protecting freedom of speech.

Individuals who choose *both* alternative *b* (Giving people more say in important governmental decisions) and alternative *d* (Protecting freedom of speech) were classified as *postmaterialists*, while those choosing both *a* (Maintaining order in the nation) and *c* (Fighting rising prices) were labeled *materialists*. Other combinations are described as *mixed.* The percentage of respondents in four western states (California, Oregon, Washington, and Idaho) choosing each of the values as first or second among the list of four is shown in Table 8.4.

Table 8.4 Percentage choosing values as first or second among four

	Total		California		Oregon		Washington		Idaho	
	1st	2nd	1st	2nd	1st	2nd	1st	2nd	1st	2nd
Maintain order	27%	18%	30%	19%	27%	19%	25%	16%	25%	20%
People more say	38%	30%	36%	29%	39%	32%	41%	31%	38%	30%
Rising prices	15%	26%	13%	27%	14%	25%	16%	26%	16%	26%
Protect speech	20%	26%	21%	26%	20%	24%	19%	28%	21%	24%
N =	3431	3396	671	665	733	724	692	685	668	664

[a]Some totals do not equal 100% due to rounding error

Table 8.5 Distribution of postmaterial value types

	Total	California	Oregon	Washington	Idaho
Materialist	13.9%	14.3%	13.6%	12.0%	13.9%
Mixed	58.4%	59.3%	59.8%	58.2%	58.4%
Postmaterialist	27.7%	26.4%	26.6%	29.8%	27.7%
%Total[a]=	100	100	100	99.9	100
N=	2851	693	756	715	687

[a]Some totals do not equal 100% due to rounding error

As the table shows, "Giving people more say in important governmental decisions" was clearly the highest priority of the four. Across the total sample (aggregated across the four states) and for each of the four state samples, "giving people more say" was *both* the most preferred value priority and the second-most preferred value priority. Given that similarity, there were some small variations among the state samples—in California 65% chose it first or second; in Idaho the figure was 68%; and, in Oregon and Washington, the percentages were 71% and 72%, respectively. Priority preferences for the other three values generally were in the 40+ combined percentage range. In California and Oregon, however, maintaining order and protecting speech were seen as more important than fighting rising prices, while in Washington, free speech numbers superseded those of fighting rising prices and maintaining order; in Idaho, maintaining order was preferred by a slightly greater percentage than for prices and speech.

Combining the value priorities at the individual level into the two pure types (materialist and postmaterialist) produced the distributions shown in Table 8.5. Recall that materialists are those respondents who choose *both* maintaining order and fighting high prices as their top two priorities, while postmaterialists choose *both* giving people more say in government decisions and protecting free speech. The remainder of the respondents are grouped into a mixed category. As Table 8.5 shows, approximately twice as many respondents were postmaterialists as were materialists, and nearly three-fifths were mixed. There was some evidence that the number of postmaterialists has grown over recent decades. A study of Washington State respondents in the 1980s identified 21% as materialist and 19% postmaterialist, compared to the 12% and 30% figures, respectively, in Table 8.5.

As a backdrop for the analysis below, it is important to understand the extent to which these postmaterial values are associated with demographic/background attributes of citizens, as well as how those relationships vary among the states. As noted at the beginning of the chapter, postmaterial values have been widely held to account for the rise in environmental concern and activism since the 1950s. Moreover, if the relationships between postmaterial values and background vary among the four states in this study, the politics and conflict over siting may also take different paths and may be conducted in a different content. Table 8.6 contains the results of a correlation analysis (r) addressing that question.

Table 8.6 reveals some interesting differences in the demographic composition of the various postmaterial value types. Thus, while greater education attainment

Table 8.6 Correlation (r) of background variables with postmaterial values

	Total	CA	OR	WA	ID
Education	.13**	,09*	.18**	.11*	.13**
Length of residence	−.09**	−.12**	−.11**	−.02	−.09**
Employment	−.01	.01	−.05	.06	−.01
Income	−.02	−.06	−.01	.01	−.02
Gender	.03	.00	.08*	.00	.03
Age	−.05*	−.06	−.06	−.14**	−.05*

*p ≤ .05; **p ≤ .01 (1-tailed); N = 2759–2845

Table 8.7 Average NEP (New Ecological Paradigm) scores within postmaterial value groups[a]

	Total	California	Oregon	Washington	Idaho
Materialist	19.9	20.1	20.3	19.5	19.7
Mixed	20.8	21.0	20.9	21.3	20.1
Postmaterialist	22.0	22.4	22.8	21.3	21.6
N=	2835	690	751	709	685
F-Test	24.20***	7.87***	11.60***	3.97*	5.80**

*$p \leq .05$; **$p \leq .01$; ***$p \leq .001$

[a]Chapter 6 describes the construction of the New Ecological Paradigm (NEP), scores which have a possible range of 5 (lowest) to 30 (highest)

had a statistically significant relationship to the holding of postmaterial values, that relationship differed considerably among the states. That relationship was clearly stronger in Oregon than in any of the other state samples, and it was somewhat weaker in California than in the other states. Length of residence in the state was negatively related to holding postmaterial values in each of the states and overall, except for Washington. Finally, in Washington, younger residents were more likely than older residents to hold postmaterial values. Because of the much larger number of cases, a statistically significant relationship was also obtained for the overall sample, but the absolute level of the relationship was weaker.

Again, if postmaterial values drive environmental concern, and if that concern has a significant independent effect on attitudes about the siting of alternative energy technologies when they impose some negative effects on the environment (tradeoffs), and if those attitudes are more likely to be found in distinct demographic locations, then they may structure the nature of the political discourse. That is, an opposition to an AET siting that is value-based and primarily found among the more highly educated, the younger age groups, and residents with a shorter length of residence, may produce a culture clash associated with those attributes.

Given the decades-long association of postmaterialism with environmental concern, it is appropriate to see whether this relationship emerges in this study's samples, and then to uncover whether attitudes about the environmental/technology trade-offs are more likely rooted in one's background, in fundamental values, or in environmental concern, and, finally, whether postmaterialism distinguishes environmental localists from environmental globalists. Table 8.7 begins this search by

Table 8.8 Analysis of variance of AET mean tradeoff indicators among postmaterial value types[a]

	Materialist	Mixed	Postmaterialist	F-test	Sig.
Solar	2.84	2.86	2.83	.16	.850
Wind	2.70	2.71	2.91	8.88	.000
Wave	3.30	3.41	3.54	3.54	.001
Hydro	2.94	2.85	2.95	2.48	.001
Index	11.77	11.83	12.22	4.99	.007

N = 2754–2783

[a]Higher tradeoff value scores indicate greater support for protecting the environment when alternative energy technologies are said to have negative effects. The scale for the individual AET types runs from 1 (go ahead with the AET) to 5 (protect the environment)

presenting the average NEP score within each of the three postmaterial value groups (materialist, postmaterialist, mixed), overall and within each state; NEP scores were significantly higher among postmaterialists than those with mixed and materialist values. The one minor exception was in Washington, where NEP values for the postmaterialists and mixed value types were the same, although both of those were substantially higher than for the materialist value types. Moreover, when looking at state differences among those classified as holding the same value sets, postmaterialists in California and Oregon had greater levels of NEP values than did those respondents in Washington and Idaho. Likewise, even when viewing the NEP concerns of materialists, Californians and Oregonians exhibited higher levels of NEP values than did those in Washington and Idaho.

Two more questions remain in this particular analysis:

1. Are the public's views of the tradeoffs in alternative energy technology siting linked to their postmaterial value sets, even when taking into account their position on the New Ecological Paradigm and their background attributes?
2. Are those patterns affected by the particular state in which the individual resides?

Recall that Chap. 7 found that the NEP (environmental concerns) was a strong predictor of AET choices, even also while significant numbers of high NEP respondents were at the pro-AET pole of the AET Tradeoff scale. And, the above analysis has suggested that NEP values themselves seem at least partially rooted in postmaterial values. Table 8.8 and Table 8.9 extend this analysis. First, Table 8.8 shows the results of an analysis of variance of the tradeoff index and individual item means among the three postmaterial value groups. Second, in Table 8.9, the tradeoff index and each of the four AET means are regressed on all of the six background variables, plus the NEP and dummy variables for three of the four states.

The results shown in Table 8.8 reveal that statistically significant differences were obtained among postmaterial value types, relative to their scores on the overall AET tradeoff index; those differences were much greater between the postmaterialists and the other two types. Moreover, the patterns with regard to the individual types of alternative energy technologies varied substantially. On solar power, the differences among postmaterial value types were trivial when respondents were asked to assess how they felt in regard to installing a "solar power farm in a desert

Table 8.9 Regression of AET tradeoff measures on NEP, postmaterialism, state dummy variables and demographic/background variables

	Solar	Wind	Wave	Hydro	Index
	b	*b*	*b*	*b*	*b*
NEP	.27***	.23***	.29***	.38***	.43***
Postmaterial	−.03	.04*	.03	−.02	.01
Length of residence	.00	.02	.00	−.09	−.04
Education	−.04	.06*	.05*	−.01	.02
Gender	−.02	.02	−.10***	.10***	.07***
Age	.01	.09***	.02	.08**	.07**
Employed	−.11***	−.02	−.03	−.07**	−.09**
Income	−.08**	.06**	.02	.03	−.02
CA dummy	−.02	−.04	.03	.00	−.01
OR dummy	.03	.00	−.02	.00	.01
WA dummy	.02	−.03	.02	.00	−.01
Adj. R2=	.10	.06	.11	.18	.21
F=	24.2***	15.7***	29.8***	48.9***	61.1***
N=	2543	2530	2539	2538	2530

*$p \leq .05$; **$p \leq .01$; ***$p \leq .001$

even if it disrupts ecology of endangered turtle." When asked whether to "continue to operate hydroelectric dams for power even if it harms salmon runs," the mixed value types were *less* likely to support the protection of salmon than were *either* the materialists or the postmaterialists. On the other hand, postmaterialists were significantly *more* likely to oppose a wind energy farm if it damaged "tall grass prairie ecology" and the installation of off-shore wave turbines on the ocean floor if the installation "impedes whale migration."

Table 8.9 shows that, while controlling for demographic variables, save for a single case (that of wind), the effect of postmaterialism on positions on the AET tradeoff issues was insignificant when taking into account an individuals' positions on NEP. Moreover, even for wind power, the effects of postmaterialism were very small. This suggests that while postmaterialism may be one source of the NEP, there are probably others, and that mixed causal ordering produces an NEP that structures AET views, regardless of the NEP source (postmaterial or otherwise). Perhaps the forces that produced the cycle that led to postmaterial values growth and influence on relevant policy questions—such as those dealing with environmental concern—have now been overtaken by other forces. Some to consider may include the worldwide economic crisis of the early twenty-fist century, the seeming realities of carbon-based climate change, the perception of significant declines in energy-based natural resources, and the trend toward viewing a simpler life as a better life. Each of those, as well as other forces, could lead environmentalists toward opposing AET even without the buttressing of postmaterial values.

To be sure, the results show that there were also environmentalists who supported AET growth, even in the face of negative environmental consequences, perhaps as a result of a larger picture. In any case, it is clear that those with greater levels of environmental concern were more likely to oppose AETs with a negative environmental impact, regardless of type of AET, state of residence, and a whole range of personal attributes.

At the same time, there were personal characteristics of individuals that retained their impact on AET opinions, even when controlling for NEP and P-Mat attitudes. Gender, age, and employment status retained significant effects on the overall AET tradeoff index, although the magnitude of the effects was substantially smaller than for the NEP. Thus, women, older individuals, and those employed fulltime were more likely to oppose AET installation when it had the potential for further environmental damage. The effect on the overall index (as well as the individual AET modes) of the state of residence was nil when the other variables in the equation were controlled.

Some intriguing contrasts appear when comparing different types of AET. Higher income respondents were more likely to reject wind energy installations with negative environmental effects, whereas they were less likely to reject solar AET installations under the same conditions. Women were more likely to reject wave and hydro AETs with negative environmental effects than men were, but the same was not the case for wind and solar installations. And, only for hydropower did length of residence have a significant effect—the longer a respondent had lived in a particular state, the less likely he or she was to oppose an AET if it had a negative environmental effect. Perhaps that finding results from respondents becoming somewhat inured to those effects after long-term familiarity.

Hydropower exhibited the effects of greater numbers of independent variables than any of the other technologies. Thus, as noted earlier, women and older respondents were more likely to reject hydropower that had negative effects, as were those who were newer to a region, and those who were fully employed. The long decades of conflict over hydropower in the West, its centrality to the Northwest power grid, the recent movement to remove dams that impede anadromous fisheries, the centrality of salmon to Pacific Northwest culture—these and other forces may have served to allow citizens to sort out their feelings in a more systematic way.

One further question requires answering, namely this: Do the *localists* (high-NEP/low-AET) and the *globalists* (high-NEP/High AET) differ in their postmaterial values? The answer is "yes," but only slightly, and that slight difference is somewhat of a surprise. High NEP/Pro-AET respondents (the globalists) were slightly more postmaterial (average of .42) than were the High NEP/low AET supporters (average of .38). While not to make too much out of a relatively small difference, it may stem from the connection of the global environmental reach of the sources of the move toward AETs, the local reach of the negative environmental impact of the installations, and the broad, international context for the development of postindustrialism.

8.5 Conclusion

This chapter has identified environmental trends and policies in postindustrial nations. In general, population growth rates are either constant or declining, there has been a dramatic increase in environmental awareness, and there is a widespread willingness on the part of individual citizens and governments to take appropriate action to clean up and protect the air, water, and landscapes of postindustrial countries. However, while these societies are increasingly "green" in outlook and in public policy, consumption rates of natural resources remain high, and the impact of postindustrial lifestyles on global forests, seas, and the atmosphere seems to place the future of the planet *and* postindustrial lifestyles at some continuing risk. This paradox characterizes the environmental policy process in most postindustrial nations. We want a clean environment and we are willing to change some of our behaviors, such as participating in household and yard debris recycling. Yet, at the same time individuals also are very much wedded to a materialist lifestyle—one that produces significant pollution and requires the continued high rate of consumption of the world's natural resources.

Postmaterialists were more likely to oppose several AETs when damages to local environments are identified. Yet, such was not the case for several other AETs. Thus, perhaps postindustrial societies are currently in a paradigm transition from the Dominant Social Paradigm, which emphasizes economic growth and resource consumption, to the New Environmental Paradigm, the emphasis of which falls on environmental concerns and resource conservation. Yet, one cannot be complacent that those holding NEP or postmaterial values will be in consensus in their support for AETs, or, for that matter, the mitigation of locally focused negative environmental impacts of those technologies. Or perhaps there is hope that current rates of growth and consumption can be maintained through scientific and technological breakthroughs. Whatever the cause of this paradox, environmental politics and policy in the postindustrial nations will remain a process marked by both conflict and controversy in the foreseeable future, including in attitudes about alternative energy technologies.

References

Aaron, H., Mann, T. E., & Taylor, T. (Eds.). (1994). *Values and public policy*. Washington, DC: The Brookings Institution.

Abramson, P., & Inglehart, R. (1995). *Value change in global perspective*. Ann Arbor: University of Michigan Press.

Ansolabehere, S., & David Konisky, D. (2014). *Cheap and clean: How Americans think about energy in the age of global warming*. Cambridge, MA: MIT Press.

Boggs, C. (2000). *The end of politics: Corporate power and the decline of the public sphere*. New York: Guilford Press.

Bryner, G. (1997). *From promises to performance*. New York: W.W. Norton.

Cada, G., Ahlgrimm, J., Bahleda, M., Bigford, T., Stavrakas, S. D., & Hall, D. (2007). Potential impacts of hydrokinetic and wave energy conversion technologies on aquatic environments. *Fisheries, 32*(4), 174–181.

Caldwell, L. (1992). Globalizing environmentalism: Threshold of a new phase in international relations. In R. Dunlap & A. Mertig (Eds.), *American environmentalism*. Philadelphia: Taylor and Francis.

Corner, A., & Randall, A. (2011). Selling climate change: The limitations of social marketing as a strategy for climate change public engagement. *Global Climate Change, 21*, 1005–1014.

Dalton, R. (1988). *Citizen politics in western democracies: Public opinion and political parties in the United States, Great Britain, West Germany and France*. Chatham: Chatham House.

Devine-Wright, P. (2010). *Renewable energy and the public: From NIMBY to participation*. New York: Routledge.

Devine-Wright, P. (2011). Place attachment and public acceptance of renewable energy: A tidal energy case study. *Journal of Environmental Psychology, 31*, 336–343.

Dunlap, R. E., Van Liere, K., Mertig, A., & Jones, R. (2000). Measuring endorsement of the new environmental paradigm: A revisited NEP scale. *Journal of Social Issues, 56*, 425–442.

Ferris, D. (2012). Geothermal company drills into a Volcano. *Forbes*. www.forbes.com/sites/davidferris/2012/10/08/geothermal-company-drills-into-a-volcano. Accessed 6 Oct 2013.

Fischer, A., Peters, V., Vavra, J., Neebe, M., & Megyesi, B. (2011). Energy use, climate change and folk psychology: Does sustainability have a chance? Results from a qualitative study in five European countries. *Global Environmental Change, 21*, 1025–1034.

Flanagan, S. (1982). Changing values in postindustrial society. *Comparative Political Studies, 14*, 99–128.

Gallup Poll. (2010). *Environment*. http://www.gallup.com/poll/1615/environment.aspx. Accessed 15 Aug 2016.

Gallup Poll. (2015). *In U.S., concern about environmental threats eases*. http://www.gallup.com/poll/182105/concern-environmental-threats-eases.aspx. Accessed 15 Aug 2016.

Habermas, J. (1981). New social movements. *Telos, 49*, 33–37.

Haggett, C. (2011). Understanding public responses to offshore wind power. *Energy Policy, 39*, 503–510.

Haggett, C. (2012). The social experience of noise from wind farms. In J. Szarka, R. Cowell, G. Ellis, P. Strachan, & C. Warren (Eds.), *Learning from wind power: Governance, societal and policy perspectives*. Palgrave Macmillan: London.

Huntington, S. (1974). Postindustrial politics: How benign will it be? *Comparative Politics, 6*, 147–177.

Inglehart, R. (1991). *Culture shift in postindustrial society*. Princeton: Princeton University Press.

Inglehart, R. (1995). Public support for environmental protection: Objective problems and subjective values in 43 societies. *Political Science and Politics, 28*, 57–72.

Inglehart, R. (1997). *Modernization and postmodernization: Cultural, economic, and political change in 43 societies*. Princeton: Princeton University Press.

Inglehart, R., & Baker, W. (1999). Modernization, cultural change, and the persistence of traditional values. *American Sociological Review, 65*, 19–51.

Johnson, G., Erickson, D., Strickland, W. P., Strickland, M. D., Shepherd, M. F., Shepard, D. A., & Sarapro, S. A. (2003). Mortality of bats at large-scale wind power development at Buffalo Ridge, Minnesota. *The American Midland Naturalist, 150*, 332–342.

Kikuchi, R. (2008). Adverse impacts of wind power generation on collision behaviour of birds and anti-predator behaviour of squirrels. *Journal for Nature Conservation, 16*, 44–55.

McCright, A. M., & Dunlap, R. E. (2011). Cool dudes: The denial of climate change among conservative white males in the United States. *Global Climate Change, 21*, 1163–1173.

Milbrath, L. (1984). *Environmentalists: Vanguard for a new society*. Albany: SUNY Press.

Offe, C. (1984). New social movements: Challenging the boundaries of institutional politics. *Social Research, 52*, 817–868.

Pasqualetti, M. J. (2000). Morality, space, and the power of wind-energy landscapes. *Geographical Review, 90*(3), 381–394. doi:10.1111/j.1931-0846.2000.tb00343.x.

Pasqualetti, M. J. (2011). Opposing wind energy landscapes: A search for common cause. *Annals of the Association of American Geographers, 101*(4), 907–917. doi:10.1080/00045608.2011.568879.

Petrova, M. A. (2013). NIMBYism revisited: Public acceptance of wind energy in the United States. *Wiley Interdisciplinary Reviews: Climate Change, 4*(6), 575–601. doi:10.1002/wcc.250.

Pierce, J. C., Steger, M. A., Steel, B. S., & Lovrich, N. P. (1992). *Citizens, political communication and interest groups: A study of environmental organizations in Canada and the United States*. New York: Praeger Publishers.

Pierce, J. C., Lovrich, N. P., Steel, B. S., & Steger, M. A. (2000). *Political culture and public policy in Canada and the United States: Only a border apart?* New York: Edwin Mellen Publishers.

Robinson, J. (2004). Squaring the circle? Some thoughts on the idea of sustainable development. *Ecological Economics, 48*, 369–384.

Simon, C. (2007). *Alternate energy: Political, economic and social feasibility*. New York: Rowman and Littlefield.

Slattery, M. C., Johnson, B. L., Swofford, J. A., & Pasqualetti, M. J. (2012). The predominance of economic development in the support for large-scale wind farms in the US Great Plains. *Renewable and Sustainable Energy Reviews, 16*(6), 3690–3701.

Steel, B. S., Pierce, J. C., Warner, R. L., & Lovrich, N. P. (2015). Environmental value considerations in public attitudes about alternative energy development in Oregon and Washington. *Environmental Management, 55*, 634–645.

Strachan, P. A., & Lal, D. (2004). Wind energy policy, planning and management practice in the UK: hot air or a gathering storm? *Regional Studies, 38*(5), 549–569.

Wolsink, M. (2010). Near-shore wind power — Protected seascapes, environmentalists' attitudes, and the technocratic planning perspective. *Land Use Policy, 27*(2), 195–203.

Woody, T. (2010, February 1). It's green against green in Mojave Desert solar battle, February 2010. *Environment, 360*. http://e360.yale.edu/feature/its_green_against_green_in_mojave_desert_solar_battle/2236/. Accessed 22 Nov 2015.

Yankelovich, D. (1989). How changes in the economy are shaping American values. In H. Aaron, T. Mann, & T. Taylor (Eds.), *Values and public policy*. The Brookings Institution: Washington, DC.

Chapter 9
The Devil You Know

Abstract This chapter focuses on the intersection of AET tradeoff preferences (i.e., local environmental versus global warming considerations), political orientations, climate change beliefs, and political efficacy. The chapter will begin with a discussion of the role of political orientations in shaping views about climate change and climate change science—in particular, the claims of some that opposition to AET as a replacement for reduced carbon-based fuel use is a political position and not one associated with environmental attitudes themselves. This chapter's findings suggest that indeed political orientations—more specifically liberal/conservative political ideology--are an important predictor of AET tradeoffs, but fail to erase the impact of the New Ecological Paradigm.

Keywords Climate change beliefs • Political ideology • Postmaterial values • New Ecological Paradigm • AET value tradeoffs

9.1 Introduction

> *The radical left loves attacking people as anti-science when anyone dares question their computer models on global warming.* (Senator Ted Cruz (Foran 2015))
>
> *Look, if anybody still wants to dispute the science around climate change, have at it. You'll be pretty lonely, because you'll be debating our military, most of America's business leaders, the majority of the American people, almost the entire scientific community and 200 nations around the world who agree it's a problem and intend to solve it.* (President Barrack Obama (Dokoupil 2016))

Recent decades have seen a growing polarization between the ideological left and the ideological right regarding issues of science, including, but not limited to, climate change beliefs (McCright and Dunlap 2011a). Mooney (2005) suggests that the extent of this divide amounts to a partisan "war on science," with conservative Republicans outright denying the general scientific consensus on issues ranging from climate change to evolution. A 2012 study found that ideological conservatives and moderates are "less confident in science than liberals" (Gauchat 2012, p. 177). Further, among conservatives, trust in science has declined since the 1970s, but has remained relatively stable for liberals and moderates (Gauchat 2012; Lewandowsky et al. 2013). A further distinction concerns lower levels of trust in

J.C. Pierce, B.S. Steel, *Prospects for Alternative Energy Development in the U.S. West*, Environmental Challenges and Solutions 4,
DOI 10.1007/978-3-319-53414-5_9

science, while endorsing free markets. On issues like climate change, "people who embrace a laissez-faire vision of the free market are less likely to accept that anthropogenic greenhouse gas emissions are warming the planet than people with an egalitarian-communitarian outlook" (Lewandowsky et al. 2013, p. 1).

Conversely, those on the left have been generally more likely to embrace science (Steel et al. 2006) and more specifically climate change science (McCright and Dunlap 2011a, b; Marquart-Pyatt et al. 2014). And, according to Anthony Giddens, the issue of climate change "...offers the opportunity to renew the case against markets that has for so long been associated with left-of-centre traditions" (2011, p. 48). Giddens further argues that ideology plays a central role in the climate change debate (2011, p. 49):

> There is...a left/right tinge to current climate change debates: those who want to respond to climate change through widespread social reform mostly tend towards the political left; most of the authors who doubt that climate change is caused by human agency, on the other hand, are on the right.

This chapter focuses on the intersection of AET tradeoff preferences (i.e., local environmental versus global warming considerations), political orientations, climate change beliefs, and political efficacy. The role of political orientations in shaping views about climate change and climate change science will be examined—in particular, the claims of some that opposition to AET as a replacement for reduced carbon-based fuel use is a political position and not one associated with environmental attitudes themselves. Then the analysis will turn to how these political views impact support or opposition to AET as a replacement for reduced carbon-based fuels, while controlling for how the public perceives their role in the AET policy process. The chapter will begin with brief reviews of the literature on ideology and climate change beliefs, ideology and support or opposition to AET, and finally a brief discussion of political efficacy/procedural justice and support for AET development.

9.2 Ideology, Partisanship, and Climate Change Beliefs

As discussed above, the academic literature on political ideology and climate change beliefs has fairly consistently found that those on the ideological right and Republicans are significantly more likely to doubt that climate change is happening and that it is caused by human activities. McCright and Dunlap analyzed Gallup Polls between 2001 and 2010 and found "...strong evidence of both ideological and partisan polarization on global warming beliefs and concern over the past decade" (2011a, p. 179). More specifically, they found increasing polarization over the climate change issue, with conservatives and Republicans in general becoming more skeptical of climate change, compared to liberals and Democrats. In fact, McCright and Dunlap suggest that it is becoming increasingly difficult to adopt effective climate policy at the national level due to the rise and influence of the Tea Party, which adopted a climate change "litmus test" for candidates running in the 2010 election (2011a, p. 179).

In another study by McCright and Dunlap (2011b), the characteristics of "climate deniers" were examined by using Gallup Surveys from 2001 to 2010. The results of this longitudinal study found that ideology plays a strong role in climate change skepticism—especially among white males (2011b, 1171):

> We do find strong evidence for a conservative white male effect on climate change denialism, whereby conservative white males are more likely than other adults to espouse climate change denial. Further, we find that those conservative white males who self-report understanding global warming very well-confident conservative white males—express an even greater degree of climate change denial.

The results of this study are consistent with the Pew Research Center's public polls over the years as well (Kiley 2015). Examining responses to climate change questions from surveys conducted over 2006 to 2015, partisanship and ideology were found to be the strongest predictors of climate change beliefs. In fact, for the 2015 survey, Pew found that 92% of liberal Democrats agreed with the statement "there is solid evidence the Earth's average temperature is rising," compared to just 38% of conservative Republicans (Kiley 2015, p. 1).

Many other studies have come to the same conclusion, namely that ideology does play a significant role in shaping climate change beliefs, even when controlling for a variety of demographic characteristics (e.g., Hamilton 2008; Wood and Vedlitz 2007). Additional research has found that "conservative ideologues are less concerned about climate change science because they believe scientists do not have adequate/clear understanding of climate change science" (Zia and Todd 2010, p. 14). Clearly ideology is an important factor in shaping the public's beliefs concerning climate change, which may well manifest itself in acceptance or opposition to AET. That relationship is discussed next.

9.3 Ideology and AET

In 2012, the Obama Administration proposed to Congress an alternative (renewable) energy program for the U.S. military, with the U.S. Navy taking a very active role promoting the policy. The Navy and the Obama Administration proposed the program as a way to promote the U.S. renewable energy industry, but also to develop advanced biofuels for the military in order to reduce reliance on foreign sources (Carswell 2012). This led Senator James Inhofe (R-Oklahoma)—an ardent global warming skeptic and member of the Armed Services Committee—to state (U.S. Senate 2012):

> I have had growing concerns about (how) President Obama's global warming agenda is harming the military. In reality, it is President Obama's war on affordable energy that is having a dramatic impact on our national security—a war that is further depleting an already stretched military budget and putting our troops at risk.

This has been a pattern in Congress, with conservative Republican Senators and House members often opposing renewable energy legislation and questioning global warming science. Yale Law School Professor E. Donald Elliot has written on the reasons why the U.S. as a whole and conservatives more specifically oppose renewable energy policy (Elliott 2013):

> [W]e have a very strong free market ideology in the United States…Many of our leading conservative think tanks in the United States, such as the Cato Institute, the Heritage Foundation, and the American Enterprise Institute, are attacking the concept of government promoting green energy and a green energy future. It is not so much that they are opposed to renewable energy per se, but rather to the heavy hand of government to mandate it.

Similarly, two studies that examined the relationships between ideology, partisanship and the adoption of renewable energy policies at the state level also found that more conservative states were less likely to adopt AET policies. Delmas and Montes-Sancho found that the percentage of Democrats in legislatures was significantly and positively associated with the adoption of Renewable Portfolio Standards (RPS) and Mandatory Green Power Options (MGPO) when compared to states with higher percentages of Republicans (2011). Hess et al. also found that between 2004 to 2014, liberal "Blue" states were significantly more likely than conservative "Red" states to adopt renewable energy and energy efficiency legislation (Hess et al. 2016). They also found that when conservative states do adopt legislation promoting renewable energy options, market-based approaches such as tax cuts and regulatory reductions were most likely used.

While there is not much research concerning the relationship between support for AETs and political ideology at the individual level, there are several studies that may be informative. Gromet et al. (2013) found that politically conservative individuals were much less likely than liberals to support investments in energy efficient technologies or to purchase energy efficient technologies (e.g., light bulbs) when labeled as environmentally friendly. These findings led the authors to conclude that *how renewable energy policies are framed* is of utmost importance in an ideologically polarized environment like the U.S. Based on their findings they conclude (Gromet et al. 2013, p. 9314):

> Therefore, appeals regarding energy efficiency that label these choices as reflecting concern for the environment might repel a substantial segment of the U.S. population that does not identify as politically liberal.

Two other political value orientations to be included in the following analysis of AET tradeoffs were the subject of previous chapters—the NEP and postmaterialist values. The previous analysis found that the NEP and postmaterialist values are strong predictors of support for the development of renewable energy. These findings correspond to previous research conducted in the states of Washington and Oregon concerning public support or opposition to the development of AETs (Steel et al. 2015). General public surveys found that those "respondents with higher NEP scores are more supportive of government promotion of wind, geothermal and solar technologies than are those with lower NEP scores" (2015, p. 642). This study also found that those respondents with postmaterialist values were significantly more likely than those with materialist or mixed values to support AET *and* to be supportive of environmental issues. This chapter now examines the effect of these and other political orientations on AET tradeoffs.

9.4 Political Efficacy

A large body of recent literature suggests that AET developers and policy makers can facilitate community trust through transparent, collaborative policy processes that engage stakeholders at all stages of the AET development and siting processes (Lofstedt 1999; Gross 2007; Wolsink 2007; Hagget 2010). Such participatory decision-making allows policy makers and developers to familiarize themselves with community values and help address community concerns in a proactive manner. Research concerning siting of AETs, such as large-scale wind farms, has focused heavily on the importance of political efficacy, often labeled "procedural justice," in creating an atmosphere of transparency, fairness, and openess that can garner community support (Wolsink 2007; Gross 2007; Devine-Wright 2012). There is also evidence that individuals who feel that their participation in AET siting issues is valued will seek even greater levels of involvement in renewable energy developments (Devine-Wright 2005).

Additional research by Kellstedt et al. concerning personal efficacy and climate change concern found that individuals "...who feel personally responsible for climate change show far greater concern for the future effects of global warming and climate change" (Kellstedt et al. 2008, p. 120–121). Therefore, personal efficacy concerning climate change is related to support for activities to mitigate future effects. So taken together, political and personal efficacy have positive impacts on individual involvement in community based AET siting, in concern for long-term effects of climate change, and in supporting activities to mitigate those effects. The analysis next turns to the public survey data from California, Oregon, Washington and Idaho to examine the relationships between ideology, climate change belief, and political efficacy.

9.5 Analyses

The first question to be addressed is whether political ideology has a simple, first-order relationship with attitudes about the tradeoffs between installing AETs and the possible environmental risks inherent in that installation. As discussed above, this quest enters the realm of political ideology because of the large literature that suggests that climate change skepticism is rooted in conservative beliefs and organizations. Thus, one would expect that skepticism to also structure attitudes about the development and installation of AETs designed to mitigate climate change. Many conservatives are said to oppose public subsidies for alternative energy

Table 9.1 Distribution of self-identified political orientation by state and overall

	California	Oregon	Washington	Idaho	Total
Very liberal	9%	7%	6%	4%	6%
Liberal	20	23	24	20	22
Moderate	44	44	45	30	41
Conservative	23	22	20	32	24
Very Conservative	4	5	5	13	7
Total[a]	100%	101%	100%	99%	100%
N=	689	748	705	687	2829

[a]Totals may not equal 100% because of rounding error

technologies, as well as requirements by states for increasing the percentages of energy production to be based in non-carbon based renewable energy forms. The proposals for those technologies and goals are often seen as a "liberal" or "environmentalist" agenda. This study relies on a widely used self-identification indicator of political ideology:

On domestic policy issues, would you consider yourself to be?

(1). Very Liberal (2). Liberal (3). Moderate (4). Conservative (5). Very Conservative

The distribution of the self-identified political orientations for the entire sample, and for each state's sample, is shown in Table 5.1

As Table 9.1 shows, the Idaho public differs substantially from the other three states by being much more conservative. In Idaho, 45% of the respondents self-identified as conservative or very conservative. This compares with figures ranging between 25% and 27% for California, Oregon, and Washington. Thus, given the research cited above, one would expect Idahoans to be less likely to believe that there is "solid evidence that the average temperature on earth has been getting warmer over the past few decades." That expectation is confirmed in these data, but possibly not as strongly as one might expect. Sixty-one percent of the Idaho respondents agreed with that statement, compared to 69%, 71%, and 72% in California, Oregon, and Washington respectively.

So, the question emerges as to whether Idaho's somewhat greater climate change denial (26% v. 17%, 17%, and 19% in the other states) is simply a consequence of the greater conservatism among its residents. We address this question in Table 9.2, which shows the percent believing the earth is getting warmer by self-placement on the political ideology scale. What is most striking is that only 7.7% of Idahoans who are "very conservative" believe the earth is getting warmer, in comparison to 44% in California, 52.9% in Oregon, and 28.9% in Washington.

In each of the four states, the very liberal respondents were much more likely than the very conservative respondents to believe that climate warming is real. At the same time, though, those differences were much bigger in Washington and Idaho than in California and Oregon. That pattern is largely the result of the substantial *disbelief* about climate warming found among the very conservative respondents in Idaho and Washington, as compared to those in California and Oregon. This

Table 9.2 Belief in climate change by state[a]

Political orientation	California	Oregon	Washington	Idaho
Very liberal	85.5%	90.0%	92.7%	88.5%
Liberal	84.3	91.0	89.8	90.5
Moderate	72.3	71.6	72.1	73.7
Conservative	43.8	46.1	55.7	49.3
Very conservative	44.0	52.9	28.9	07.7
N=	693	756	698	684
Chi Sq=	110.10	126.79	122.76	303.27
Sig.=	.000	.000	.000	.000

[a]The entry in each cell is the percent that there is "solid evidence that the earth is getting warmer" by political orientation

Table 9.3 Correlation (r) of AET responses with political ideology and climate change belief[a]

	Political ideology	Climate change belief
Overall index	−.27***	−.18***
Solar	−.17***	−.09***
Wind	−.10***	−.07***
Wave	−.21***	−.14***
Hydro	−.23***	−.19***

***$p \leq .001$; N = 2814—2835

[a]Negative sign means belief in climate change is associated with pro-local environment position on AET questions

latter pattern accounts for the significant effects attributed to the interaction between political orientation and state of residence.

The question now becomes whether political orientation and climate change beliefs have any relationship to responses to AET tradeoff questions. This question is answered via the results shown in Table 9.3, which reflects each of the tradeoff question responses correlated (r) separately with political orientation and climate change. Both political orientation and climate change had a significant correlation with the alternative energy tradeoff questions. That is, liberals were more likely than conservatives to oppose AET installations if the AET also had negative effects on the environment. Moreover, perhaps ironically, respondents who believed that "there is solid evidence that the average temperature on earth has been getting warmer over the past few decades" were also more likely to oppose the installation of AETs said to be designed to mitigate climate change, if those technologies also were said to have negative effects on the local environment. At the same time, those relationships for climate change beliefs were weaker than for the political orientation question. Thus, perhaps orthogonal to the global/local contrast, there appears to be some indication of a short-term/long-term tradeoff or perhaps a visible/invisible, diffuse/specific-particularistic or concrete/vague tradeoff, such that even for those concerned with climate change, the more immediate negative consequences for

Table 9.4 Correlations (r) among NEP, postmaterialism, climate change and political ideology

	Climate	Postmaterialist	NEP	Political
Climate change	X	−.09**	−.40**	.42**
Postmaterialist		X	.13**	−.27**
NEP			X	−.48**
Political ideology				X

**$p \leq .01$; N = 2814—2835

turtles or fish or grasslands or whales are the more central to their beliefs. This may be the ecological take on the old saying that "better the devil you know than the devil you don't."

The implication of the above results may be that accommodation or rejection views of the possible negative impacts of AETs may be driven more by fundamental environmental values than by those beliefs connected to the presence or absence of climate change. As a first look into that possibility, Table 9.4 shows the correlations among political orientations, climate change beliefs, postmaterial value orientations, and NEP values. The perceived climate change variable is strongly related to environmental values, as expressed in the NEP ($r = -.40$, $p \leq .001$) and to political/ideological self-identification ($r = .42$, $p \leq .001$), and weakly (albeit statistically significant) related to postmaterialist values ($r = -.09$, $p \leq .001$). Thus, those who agree that the earth is getting warmer are more likely to be liberal, strongly pro-environmental and hold postmaterialist values.

Table 9.5 takes the analysis a step further; the tradeoff beliefs are regressed on the independent variables in two models. The first model (I) uses only the four attitudinal/value questions as independent variables. The second model (II) retains those four variables and adds into the predictors the six demographic/background variables. Thus, one can observe whether the background attributes add any explanatory power to the predictive model that includes only the four attitudinal variables. Moreover, one can determine whether the simultaneous consideration of the background variables along with the attitudinal variables reduces the effects of the latter. (All models were tested for potential multicollinearity and found not to be at risk.)

When simultaneously considering all four general orientations, along with the individual background variables, it is clear that the independent variable with the most consistent and strongest effects was the New Ecological Paradigm. The NEP had significant effects ($p \leq .001$) on all five tradeoff dependent variables under both sets of models. Those with greater NEP values were more likely to reject the putative long-term climate-saving AET installations when those installations have more specific and immediate alleged negative local effects on the environment. Political orientations, in the form of liberal/conservative self-identification, had significant effects on three of the four forms of AET and the overall summative index. Liberals were generally are more likely to take the specific local environmental protection position than were conservatives, albeit the effect sizes were much smaller than was the case for the NEP. The acceptance of "solid evidence" of climate change had a small, but significant, effect on accepting solar power as an AET, even if it threatens

Table 9.5 Regression of tradeoffs on beliefs and backgrounds

	Tradeoff									
	Index		Solar		Wind		Wave		Hydro	
	I	II	I	II	I	II	I	II	I	II
	b	*b*	*b*	*b*	*b*	*b*	*b*	*b*	*b*	*b*
Political ideology	–.08***	–.09**	–.08***	–.09***	.01	–.01	–.08***	–.07**	–.05*	–.06**
Climate change	.01	.01	.04*	.03	.02	.02	.00	.00	–.03	–.02
Postmaterialist	–.02	–.01	–.05**	–.04*	.04*	.04	.02	.02	–.06**	–.04*
NEP	.41***	.40***	.27***	.24***	.22***	.24***	.25***	.26***	.37***	.34***
Age		.08**		.02		.09***		.03		.08***
Gender		–.07***		–.02		.02		–.10***		–.09***
Education		.01		–.04*		.06**		.03		–.02
Length of residence		–.04***		–.04		.02		–.01		–.09***
Employed		–.08***		–.10***		–.02		–.04		–.07**
Income		–.02		–.10***		–.02		.02		.03
R^2 =	.20	.22	.09	.10	.05	.06	.09	.11	.16	.18
F =	170.1	68.3	63.3	28.0	34.3	16.7	70.0	31.7	133.0	53.8
F Sig. =	.000	.000	.000	.000	.000	.000	.000	.000	.000	.000
N =	2726	2501	2736	2514	2728	2503	2730	2510	2730	2509

*p ≤ .05; **p ≤ .01; ***p ≤ .001

endangered turtle populations. Postmaterialist values had significant effects for solar, wind, and hydropower alternative energy technologies, although none of those effects reached the size of those for the NEP.

Model II for each of the AET indicators shows the effect of adding the background variables to the regression equation. In every case, the effects on the overall explained variation (R^2) were marginal, ranging from only 1% to 2%. The individual background characteristics had a range of significant effects. Age was related to the overall index, to the wind-power tradeoff, and to the hydropower tradeoff opinion. In each *significant* case, the older, the female, and the full-time employed respondents were more likely to oppose AET installation if it would produce immediate damage to the environment. In regard to solar power, the more highly educated and the higher income respondents were less likely to support solar installations if they have a negative impact on the environment.

Of the specific AET forms, hydropower appeared to exhibit greater effect by the independent variables than do the other AETs (R^2 of .18 in the full model as opposed to .10, .06 and .11 for the other AET forms). Moreover, even when considered simultaneously, political ideology, postmaterialist values, the New Ecological Paradigm, age, gender, length of residence, and employment were significant independent predictors of attitudes about hydropower in the face of negative near-future environmental impact. Thus, the hydropower/anadromous fisheries tradeoff attitudes clearly are bound up in the mix of political and cultural values, as well as identifiable social locations. In comparison, however, the explained variation produced by regressing tradeoff beliefs and background attributes was smaller for the wind power AET than for any of the others. Moreover, the effect size for the NEP independent variable for wind power (b = .24) was smaller than for hydropower (b = .34).

The findings continue to underscore a pattern of surprisingly strong numbers of environmentalists who oppose the installation of alternative energy technologies—that are intended to mitigate climate change, with its negative *long-term* environmental consequences—when those technologies are said to have relatively *short-term* negative environmental consequences. At the same time, while much of the discussion of climate change focuses on carbon-based/fossil fuel by-products as a source of that change, that is not the only argument posed by even those who accede to the presence of that change. In this latter group are those who believe that while climate change may be occurring, that change is a function of natural, non-human induced processes. It is possible that such distinctions as to the cause of climate change may structure individuals' orientations to the tradeoff between AET and environmental impacts.

Thus, Table 9.6 shows the average AET tradeoff scores for each combination of the New Ecological Paradigm index and the perceived climate change cause measure. The NEP is trichotomized into approximate thirds. The change source measure also has three categories: natural causes, human causes, and "don't know." As the summary measures at the bottom of the table show, the two variables explained more variation in the AET attitudes index (R^2 = .18) and for hydro power (R^2 = .15) than was the case for the solar, wind, and off-shore alternatives. The NEP showed a

Table 9.6 Average support for AET installation in combinations of NEP and perceived source of climate change[a]

NEP	Change source	N	Index	Solar	Wind	Off-shore	Hydro
Low	Human	(106)	10.6	2.2	2.5	3.3	2.6
Low	D.K.	(84)	11.1	2.5	2.5	3.3	2.7
Low	Natural	(200)	10.4	2.6	2.6	3.1	2.1
Med.	Human	(433)	12.0	2.8	2.7	3.5	3.0
Med.	D.K.	(143)	11.3	2.8	2.5	3.3	2.8
Med.	Natural	(207)	12.0	2.9	2.9	3.3	2.9
High	Human	(523)	12.8	3.0	3.2	3.9	3.6
High	D.K.	(113)	11.9	2.8	3.3	3.5	3.3
High	Natural	(63)	11.4	2.8	2.3	3.8	3.1
R^2			.18	.09	.07	.06	.15
Sig. NEP			.000	.000	.000	.000	.000
Sig. Climate change			.004	.056	.006	.060	.000
Sig. Interaction NEP/change			.000	.000	.000	.000	.000

[a]The entry in each cell is the average score on the designated tradeoff question for respondents in the particular combination of the New Ecological Paradigm and the perceived source of climate change measures. The respondents include only those who believe that climate change is occurring

significant effect across the board, while the source of the climate change variable was not significant ($p \leq .05$) for solar and off-shore. Moreover, the interaction effect of NEP and the cause variable was significant in each case. The starkest difference in the overall index averages comes from comparing Low NEP/natural causes combination (10.4) with the High NEP/human causes pair (12.8). Moreover, the High NEP/human cause set of respondents was more likely than the High NEP/natural causes pair to take the near term environmental protection alternative. Similar patterns obtain throughout the table.

As forecast in Chap. 7, further unraveling of the foundations of the politics of AET may be found in the extent to which respondents are engaged in the issue and the degree to which they believe they can influence their state's energy policy. Our surveys asked respondents these questions:

How often do you talk about issues related to renewable energy with your family, friends, or other acquaintances? The response alternatives included: (1). Never; (2). Hardly ever; (3). Sometimes; and (4). Often.

Overall, how much impact do you think PEOPLE LIKE YOU can have in making (your state's energy policy? The response alternatives included: (1). No impact at all; (2). A small impact; (3). A moderate impact; and (4). A big impact. The distributions of respondents in each state on these questions are shown in Table 9.7.

Table 9.7 shows that there were few interstate differences in the degree to which citizens talk about renewable energy, or in the degree to which they think people like them have an impact on making energy policy. Moreover, no more than 15% (California) "often" talked about renewable energy, and more than half (also California) never or hardly ever talked about renewable energy. Thus, despite the great deal of attention given to renewable/alternative energy in the media, it seems

Table 9.7 Responses to questions about whether talk about renewable energy issues, and about how much impact on energy policy, by state

		State			
		California	Oregon	Washington	Idaho
Talk about renewable energy issues	Never	11%	9%	10%	7%
	Hardly Ever	30%	29%	29%	28%
	Sometimes	44%	47%	49%	53%
	Often	15%	15%	12%	12%
	Total	100%	100%	100%	100%
	N =	693	756	714	687
	Average	2.63	2.69	2.64	2.70
	F = 1.25 Sig. = .291 N = @ 2850				
How much impact making energy policy	None	18%	16%	15%	18%
	Small	47%	46%	48%	46%
	Moderate	26%	27%	27%	25%
	A lot	9%	11%	10%	11%
	Total	100%	100%	100%	100%
	N =	692	756	712	687
	Average	2.26	2.33	2.33	2.28
	F = 1.25 Sig. = .289				

Table 9.8 Correlations (r) of talk about renewable energy and perceived impact on renewable energy policy with background and attitudes

	Talk about renewable energy	Have impact on energy policy
Climate change	−.05**	−.12**
Political ideology	−.10**	−.12**
Postmaterialist values	.14**	.06**
New Ecological Paradigm	.08**	.18**
Age	−.06**	−.13**
Gender	.10**	−.10**
Education	.13**	.02
Length in state	−.04*	−.11**
Employed	−.01	−.10**
Income	.04	.00
N =	2556	2553

*p ≤ .05; **p ≤ .01

not to have penetrated very deeply into the daily lives of most Americans. Moreover, no more than 11% of the respondents believed that people like them have a lot of impact on energy policy. Thus, it seems clear that it is a fairly small portion of the public in these western states that are engaged in the topic of alternative energy. But, are these more engaged individuals similar to the rest of the population? Table 9.8 shows for the entire sample, the correlations between the responses to the same two

Table 9.9 Correlation (r) of talk and impact with tradeoff variables

	Index	Solar	Wind	Wave	Hydro
Talk	.03	−.04*	.08**	.09**	−.03
Impact	.09**	.06**	.04	.08**	.07**

*p ≤ .05; **p ≤ .01; N = 2757–2847

questions and the sets of variables used to try to explain the alternative energy preferences, namely those of values and of demographics.

Table 9.8 shows that responses on the two engagement questions were related to the full range of the political and environmental attitudes, as well as to most of the background/demographic variables. That is, those respondents who believe that the earth is getting warmer, who self-identify as liberals, who hold postmaterialist values, and who ascribe more to the New Ecological Paradigm are more likely to report that they talk about issues related to renewable energy and that people like them have an impact on their state's energy policy. Moreover, the same patterns generally held for younger respondents. Men were more likely to report talking about renewable energy policy, but less likely to believe that people like them can have an effect on policy outcomes. More highly educated respondents were more likely to talk about renewable energy. Respondents with fewer years of living in their respective states were more likely to talk about renewable energy, and more likely to believe they could have an impact on energy policy. And, those employed full-time were more likely to believe that they have an impact on energy policy.

The next question was whether respondents who talk more about renewable energy or who think people "like them" will have a significant impact on energy policy in their particular state. In this context, Table 9.9 shows the correlation between those two engagement variables (talk and impact) and the five tradeoff variables (summary index, solar, wind, wave and hydro).

With a few exceptions, respondents who talked more about renewable energy and who felt they are more likely to have an impact were somewhat more likely to take the short-term local environmental protection side of the tradeoff questions. The exceptions were for the overall index and hydro for the talk indicator, and wind for the impact indicator. Then, a question is whether the two engagement variables retain their significant effects when controlling for the attitudinal/value variables and the demographic variables.

Table 9.10 shows the independent effects of each of the predictor variables when they are all entered simultaneously in a multiple regression equation explaining positions on the AET trade-off questions. In particular, we were interested in whether the talk and impact variables retained their significance when considering all of the other independent variables at the same time. The answer was a mixed yes. First, only the "talk about" question had a significant effect. Second, even for the talk variable, the overall index showed a negligible effect, but that seems to be because of contrasting signs on the four specific AETs.

That is, those who talked more about renewable energy were somewhat more likely to support solar energy installations, regardless of their effect on the immediate

Table 9.10 Multiple regressions of independent variables on tradeoff variables

	Index	Solar	Wind	Wave	Hydro
	b	*b*	*b*	*b*	*b*
Political ideology	−.08**	−.08***	.00	−.06*	−.07**
NEP	.40***	.23***	.23***	.25***	.35***
Postmaterialistism	−.01	−.04*	.03	.01	−.04
Climate change	−.02	.00	.00	−.02	−.03
Age	.08***	.01	.09***	.03	.08**
Gender	−.07***	−.08**	.01	−.11***	−.09***
Education	.01	−.04	.05*	.04	−.06**
Residency length	−.04*	−.03	.01	−.01	−.09***
Employed	−.08**	−.10***	−.02	.04	.03
Income	.03	−.10***	−.01	.03	.03
Talk about AET	.01	−.05*	.06**	.07***	−.07***
Effect on policy	.00	.03	−.02	−.02	.00
R^2 =	.22	.10	.07	.12	.18
F =	57.4***	23.8***	14.9***	27.9***	46.4***
N =	2509	2525	2512	2521	2520

*p ≤ .05; **p ≤ .01; ***p ≤ .001

Table 9.11 Average political belief scores for globalists and localists

Political beliefs	Globalists	Localists	
% Postmaterialist values [N = 571]	41.6%	37.7%	Chi-sq. = 2.11
Mean score for political ideology [N = 568] (1 = Very liberal to 5 = very conservative)	2.79	2.42	T-test = 12.09***
% Believe earth getting warmer [N = 485]	72.3%	88.2%	Chi-sq. = 16.92***
% Believe warming is human caused [N = 485]	64.4%	81.6%	Chi-sq. = 13.05***
Mean score for belief that people can impact energy policy [N = 569] (1 = no impact to 4 = big impact)	2.30	2.51	T-test = 4.95*
Mean score for talking about renewable issues with family, friends, etc. [N = 571] (1 = never to 4 = often)	2.60	2.84	T-test = 7.23**

*p ≤ .05; **p ≤ .01; ***p ≤ .001

environment. The same was true for the hydroelectric tradeoff question. On the other hand, for the wind and wave alternatives, the opposite was the case, namely the more the respondent was engaged in conversation about renewable energy the more likely he or she was to oppose the installation of technologies if they had negative environmental consequences.

Finally, the question must be posed as to whether in the tradeoff context, the globalists (high-NEP/high-AET support) differed from the localists (high-NEP/low AET support) on the political variables considered in this chapter. As shown in Table 9.11, globalists—who share high levels of environmental concern with the localists, but also are more willing than localists to trade off local negative environmental impacts in order to site alternative energy technologies—were less likely to be very liberal, less likely to believe that the earth's climate is getting warmer, less likely to believe that change is caused by humans, less likely to believe they have an impact on energy policy, and less likely to talk about renewable energy.

Thus, localists may not be localists because of less concern about climate change; to the contrary, localists were more concerned than were globalists about the global threat of climate change, and thereby perhaps were more broadly based in their environmental concerns and more likely take action to achieve them. And, the immediacy of the perceived negative impacts of AETs seem to have overridden the longer-term horizon of climate change.

9.6 Conclusion

The central focus of this chapter has entailed an examination of the role of political orientations and values—ideology, postmaterialist values and the NEP—in shaping citizen attitudes toward AET tradeoffs between potential local environmental considerations and the broader issue of reduction of greenhouse gases. A significant finding is that the NEP is a better predictor of the AET index scores than are the respondents' views about climate change, even though climate change is widely touted as a central reason for the development of those technologies. At the same time, paradoxically, globalists (those who choose AETs over local environmental concerns) are less likely to see the climate becoming warmer, and are less likely to see humans as the cause of that change. The localists may be more likely to perceive immediacy to the environmental consequences of the installation of an AET, in contrast to the long-term, future horizon for climate change. This chapter's findings also suggest that indeed political orientations—more specifically liberal/conservative political ideology—are an important predictor of AET tradeoffs, but fail to erase the impact of the New Ecological Paradigm.

References

Carswell, C. (2012, May 30). The ideological war against renewable energy. *High Country News*. https://www.hcn.org/blogs/goat/the-ideological-war-against-renewable-energy. Accessed 15 Aug 2016.

Delmas, M. A., & Montes-Sanch, M. J. (2011). U.S. state policies for renewable energy: Context and effectiveness. *Energy Policy, 39*, 2273–2288.

Devine-Wright, P. (2005). Local aspects of renewable energy development in the UK: Public beliefs and policy implications. *Local Environment, 10*(1), 57–69.

Devine-Wright, P. (2012). Explaining "NIMBY" objections to a power line: The role of personal, place attachment and project-related factors. *Environment and Behavior, 45*(6), 761–781.

Dokoupil, T. (2016, January 12). Obama goes big on climate change. *MSNBC*. http://www.msnbc.com/msnbc/obama-goes-big-climate-change. Accessed 15 Aug 2016.
Elliott, D.E. (2013). *Why the United States does not have a renewable energy policy*. Washington, DC: Environmental Law Institute. http://www.eli.org. Accessed 15 Aug 2016.
Foran, C. (2015, December 9). Ted Cruz turns up the heat on climate change. *The Atlantic*. http://www.theatlantic.com/politics/archive/2015/12/ted-cruz -takes-a-stand-against-science/419691/. Accessed 15 Aug 2016.
Gauchat, G. (2012). Politicization of science in the public sphere: A study of public trust in the United States, 1974 to 2010. *American Sociological Review, 77*(2), 167–187.
Giddens, A. (2011). *The politics of climate change* (2nd ed.). Cambridge: Polity Press.
Gromet, D. M., Kunreuther, H., & Larrick, R. P. (2013). Political ideology affects energy-efficiency attitudes and choices. *Proceedings of the National Academy of Sciences, 110*, 9314–9319.
Gross, C. (2007). Community perspectives of wind energy in Australia: The application of a justice and community fairness framework to increase social acceptance. *Energy Policy, 35*(5), 2727–2736.
Haggett, C. (2010). 'Planning and persuasion': Public engagement in renewable energy decision-making. In P. Devine-Wright (Ed.), *Renewable energy and the public*. Earthscan: London.
Hamilton, L. C. (2008). Who cares about polar regions? Results from a U.S. public opinion. *Arctic, Antarctic, and Alpine Research, 40*, 671–678.
Hess, D. J., Mai, Q. D., & Brown, K. P. (2016). Red states, green laws: Ideology and renewable energy legislation in the United States. *Energy Research and Social Science, 11*, 19–28.
Kellstedt, P. M., Zahran, S., & Vedlitz, A. (2008). Personal efficacy, the information environment, and attitudes toward global warming and climate change in the United States. *Risk Analysis, 28*, 113–126.
Kiley, J. (2015, June 16). Ideological divide over global warming as wide as ever. *Pew Research Center*. http://www.pewresearch.org/fact-tank/2015/06/16/ideological-divide-over-global-warming-as-wide-as-ever/. Accessed 15 Aug 2016.
Lewandowsky, S., Gignac, G. E., & Oberauer, K. (2013). The role of conspiracist ideation and worldviews in predicting rejection of science. *PloS One, 8*(10).
Lofstedt, R. E. (1999). Role of trust in the North Blackforest: An evaluation of a citizen panel project. *The Risk, 10*, 7.
Marquart-Pyatt, S. T., McCright, A. M., Dietz, T., & Dunlap, R. E. (2014). Politics eclipses climate extremes for climate change perceptions. *Global Environmental Change, 29*, 246–257.
McCright, A. M., & Dunlap, R. E. (2011a). The politicization of climate change and polarization in the American public's views of global warming. *The Sociological Quarterly, 52*, 155–194.
McCright, A. M., & Dunlap, R. E. (2011b). Cool dudes: The denial of climate change among conservative white males in the U.S. *Global Environmental Change, 21*, 1163–1172.
Mooney, C. (2005). *The Republican war on science*. Cambridge, MA: Basic Books.
Steel, B. S., Lach, D., & Satyal, V. (2006). Ideology and scientific credibility: Environmental policy in the American Pacific Northwest. *Public Understanding of Science, 15*, 481–495.
Steel, B. S., Pierce, J. C., Warner, R. L., & Lovrich, N. P. (2015). Environmental value considerations in public attitudes about alternative energy development in Oregon and Washington. *Environmental Management, 55*, 634–645.
U.S. Senate (2012, May 21). Battle over military green energy efforts heads to Senate. http://www.epw.senate.gov/public/index.cfm/in-the-news?ID=6F8C98CA-802A-23 AD-43B0-1470 F5B40DE9. Accessed 15 Aug 2016.
Wolsink, M. (2007). Planning of renewables schemes: Deliberative and fair decision-making on landscape issues instead of reproachful accusations of non-cooperation. *Energy Policy, 35*(5), 2692–2704.
Wood, B. D., & Vedlitz, A. (2007). Issue definition, information processing, and the politics of global warming. *American Journal of Political Science, 51*, 552–568.
Zia, A., & Todd, A. M. (2010). Evaluating the effects of ideology on public understanding of climate change science: How to improve communication across ideological divides? *Public Understanding of Science, 19*, 743–761.

Chapter 10
The Role of Energy Policy Beliefs

Abstract This chapter examines the role of energy policy beliefs in the generation of attitudes about the tradeoffs involved in the siting of alternative energy technologies. In this chapter we are trying to answer questions about whether perceptions of impending scarcity in the resources required for energy production, the relationship of energy production to the country's national security, and the potential ameliorative role of technology in addressing energy issues for the United States, influence beliefs about AET tradeoffs.

Keywords Energy scarcity • Energy security • Energy technology support • Foreign ownership • AET tradeoffs

10.1 Introduction

> *We've known about the dangers of our oil dependence for decades. Richard Nixon talked about freeing ourselves from dependence on foreign oil. And every President since that time has talked about freeing ourselves from dependence on foreign oil. Politicians of every stripe have promised energy independence, but that promise has so far gone unmet.* (President Barack Obama (Whitehouse 2011))
>
> *Let's expedite the regulatory and permitting and legal processes for on and offshore drilling. Instead of paying billions of dollars, hundreds of billions of dollars that now are being sent to foreign regimes, we should be drilling here and drilling now instead of relying on them to develop their resources for us. So what we've got to do is axe that plan for cap-and-tax, that policy is going to kill jobs and it's going to pass the burden of paying for it onto our working families.* (Alaska Governor Sarah Palin (On the Issues 2010))

This chapter examines the impact of energy policy beliefs on individual preferences about the tradeoffs involved in the siting of alternative energy technologies (AET). Previous research has found that public desire for energy independence from foreign sources and concern about dwindling domestic energy supplies have led to support for government programs to develop AET, such as biofuels, wind, solar, geothermal, and wave energy (Steel et al. 2015). However, little research has examined how people's energy policy beliefs may structure their views about environmental tradeoffs at the local level versus the need for AET to mitigate greenhouse gas emissions for climate change. The following analysis assesses whether anti-AET attitudes in the face of localized negative environmental consequences are

J.C. Pierce, B.S. Steel, *Prospects for Alternative Energy Development in the U.S. West*, Environmental Challenges and Solutions 4,
DOI 10.1007/978-3-319-53414-5_10

a function of attitudes and beliefs about energy policy. Those beliefs include responses to survey questions addressing energy scarcity, security, and support for energy technology development.

10.2 Prior Research

Most public opinion polling concerning energy and the environment has entailed tradeoff questions that are relatively generic. For example, the Pew Research Center for the People and the Press has used this question: "Which of the following approaches to solving the nation's energy problems do you think the U.S. should follow right now—emphasize production of more oil, gas, and coal supplies or emphasize conservation by consumers of existing energy supplies? (Pew Research Center 2015). Similarly, Gallup has also asked rather generic environment-energy tradeoff questions over the years. The following question has been included in Gallup surveys: "With which of these statements about the environment and energy production do you most agree—protection of the environment should be given priority, even at the risk of limiting the amount of energy supplies—such as oil, gas, and coal—which the United States produces or development of U.S. energy supplies—such as oil, gas and coal—should be given priority, even if the environment suffers to some extent?" (Gallup Poll 2015). While both polls show environmental concerns trending higher than energy conservation and production concerns in recent years, they do not address the specific types of AET and tradeoffs with more specific environmental concerns as investigated in this study.

Since 2011, the Pew Research Center has begun to ask a general question concerning the development of energy sources, but again the question does not address specific AET tradeoff issues between development and the environment. The Pew Research Center question is as follows: "Right now, which one of the following do you think should be the more important priority for addressing America's energy supply? Developing alternative sources such as, wind, solar and hydrogen technology; Expanding exploration and production of oil, coal and natural gas; or, both should be given equal priority" (Pew Research Center 2015). For most years, Pew has found that a majority prefers "developing alternative sources" of energy, but once again this question does not address more specific environment versus AET tradeoffs. Similarly, in *Cheap and Clean: How Americans Think about Energy in the Age of Global Warming*, Ansolabehere and Konisky found that (2014, pp. 204–205) "Most Americans tilt in the direction of the environment…our survey data reveal that most people's support for energy resources is more sensitive to perceptions of environmental harms than to economic cost." As the following section shows, the literature regarding support for AET development and public concern for energy scarcity, energy security and support for science and technology research is limited.

10.2.1 Energy Security

Most research on energy security focuses on national dependence on foreign sources of fossil fuels—most notably oil and gas—and geopolitics. Johansson has developed a typology for energy security that includes two broad themes, including energy systems "as an object exposed to security threats" and energy systems "as a subject generating or enhancing insecurity" (Johansson 2013a, p. 200). The former threat theme entails security of supply (e.g., resource availability) or security of demand (e.g., protection of market share). The latter threat includes economic and political risk factors (e.g., presence of internal corruption), technological risk factors (e.g., oil pipeline leaks), and environmental risk factors (e.g., climate change).

Johansson also argues that "energy security is sometimes used to advocate renewable energy systems," because renewable energy sources "…do not suffer from the same long-term resource availability problems as finite fossil resources and their geographical location is less concentrated" (2013b, p. 598). Similarly, the nonpartisan think tank Council on Foreign Relations conducted a public opinion survey in 2009 that asked U.S. respondents their opinions on a variety of energy security issues (2009). Survey results indicate that not only are U.S. citizens concerned about reliance on foreign sources of energy and opposed to exportation of U.S. energy supplies, but they also are supportive of the development of AET for greater energy independence (2009, p. 2):

> There is strong U.S. support for a variety of methods to address the problem of energy supply. Americans are in favor of putting greater emphasis on the development of alternative renewable sources such as solar and wind, requiring utilities to use more alternative renewable energy (even if this increases the cost), and providing tax incentives to encourage the development and use of such technologies. There is substantial optimism that investments in alternative energy will pay off economically in the long run. Americans also strongly agree that investing in renewable energy is important for the United States to remain competitive in the global economy.

Findings from a 2016 University of Texas national survey of Americans found that 58% of the public would like government and private industry to collaborate for energy security and independence reasons (University of Texas 2016). However, that poll also found partisan differences, with 66% of Democrats supporting collaborative efforts compared to 53% of Republicans. Similar findings are evident from a survey conducted for the progressive think tank Center for American Progress (2014). Respondents were asked the question, "which is the better approach for the U.S. to achieve energy independence from foreign oil?" Fifty-eight percent of Democrats preferred the response "development of renewable energy resources such as wind and solar power," and 60% of Republicans preferred "development of America's own oil, natural gas, and coal" (2014, p. 9).

10.2.2 Energy Scarcity

The recent glut of oil in world markets due to increased production in some countries—including the use of hydraulic fracking in the United States—has eased concerns about "peak oil" and energy scarcity in the United States. Indeed, Mark Perry, of the conservative think tank The American Enterprise Institute, has stated (Perry 2013) that "The shale energy revolution has turned the U.S. into an energy powerhouse. Once considered a source of vulnerability to America's superpower status, oil and gas production has now become the very foundation of the country's economic strength."

However, some analysts still believe fossil fuels will become scarce in the future with growing demand from countries such as China and India. As Nisbett et al. suggest (2011):

> Over the next few decades as society passes the point of peak production of petroleum, human health and well-being are likely to face significant risks. Despite declines in global production rates, global demand for petroleum is likely to grow. As a consequence, the price of petroleum –and overall energy costs—will begin to steadily increase over the long term.

Nisbett et al. also conducted a national public opinion survey concerning energy scarcity issues and found that 76% of respondents believed that oil prices would significantly increase over the next 5 years (2011, p. 1622). They also found that "very conservative" and "very liberal" respondents were significantly more likely to believe prices would increase when compared to ideologically moderate respondents.

A 2012 Texas A & M University national survey on energy issues also found that 60.6% of respondents believe that the U.S. is "likely" and "extremely likely" to face energy shortages in the next 10 years and 64.8% responded that the U.S. should reduce dependence on foreign sources of energy (Texas A&M University 2012). The survey also found strong public support for the development of AET, with 79.4% favoring increased use of solar energy, 76% favoring increased use of wind energy, and 64.6% favoring increased use of hydroelectric power. Gallup Polls also have asked about energy availability and affordability annually in national surveys, with the question: "How much do you personally worry about the availability and affordability of energy?" (Gallup 2016). While the percentage of people responding "a great deal" has declined from 48% in 2012 to 27% in 2016, a sizable majority of 61% responded either a "fair amount" or "a great deal" in the most recent 2016 poll. Another question Gallup asks annually is "how serious would you say the energy situation is in the United States—very serious, fairly serious, or not serious at all?" In the 2012 poll, 42% said "very serious" and 49% said "fairly serious." As with the previous question, the percentage that said "very serious" and "fairly" serious has declined to 28% and 48%, respectively, in 2016 (Gallup 2016). However, the contemporary numbers reflect that a large percentage of respondents are concerned about the U.S. energy system, and this is most likely related to support for the development of AET, as investigated below.

10.2.3 Technology Support

The final theme examined in this chapter as a predictor of AET tradeoffs considers public values concerning support for science and technology development. In recent decades, it has become clear that scientists/experts and their development of new technologies no longer hold sway as unquestioned authoritative sources of credible information in many high-stakes policy debates. The objectivity and truth claims of traditional science and corresponding technologies have come under attack by both the left (e.g., anti-vaccination) and the right (e.g., climate change) (Steel et al. 2006; Dryzek 2002).

In addition to some ideological reservations about science and technology, there are additional concerns by some environmental activists. While modern environmentalism has utilized and continues to make use of information supplied by science and scientists (e.g., Rachel Carson), there are those in the environmental movement who believe science is too often used for technologies destructive to the environment (Porwitt 2000). This critique finds its roots in the 1960s and 1970s, when many people became concerned that the widespread use of some products of high technology—such as nuclear power, powerful pesticides and fungicides, genetic engineering, etc.—posed serious threats to public health and the natural environment. A widely read example of this perspective is E.F. Schumacher's book *Small is Beautiful: Economics as if People Mattered* (1973).

In terms of support for technological development and AET, there has been little research conducted to date. One study, as discussed in the introduction to this chapter, has examined public attitudes toward government programs that promote biomass, wind, geothermal, solar, and wave energy technologies in Oregon and Washington (Steel et al. 2015). The study found that younger, more highly educated and environmentally oriented respondents were strongly supportive of government policies that promote biomass, wind, geothermal, and solar energy. Support for wave energy technology was less predictable, given the nascent nature of the technology in the region.

A mixed-methods study by Frewer et al. (1998) concerning public attitudes toward technology found that the social context surrounding technology "is likely to be one of the most important determinants of it future development and application" (p. 221). That research found that general attitudes toward technology (e.g., costs and benefits) were good predictors of support for most emerging technologies (e.g., GMOs, nuclear energy, surgery, etc.), with the exception of renewable energy such as solar. In regard to green alternative technologies such as renewable energy, the study found that "…those individuals who are not favourable towards technology in general are most likely to favour 'green' alternatives which are associated with minimal environmental impact" (1998, pp. 231–232). The authors conclude that opposition to or support for any given technology is dependent on the ratio of perceived benefits and risks to the specific technology in question.

10.3 Analyses

In order to determine the impact of energy policy beliefs on individuals' preferences among AET tradeoffs, respondents in our surveys were asked to indicate the degree to which they agreed or disagreed with the following seven statements concerning energy policy:

(a) Decreasing our dependence on foreign oil and gas is important to our national security.
(b) I am concerned about foreign ownership of our energy resources.
(c) I am concerned that our country doesn't have enough energy resources.
(d) I am concerned about being personally affected by shortage of electricity in the next 5 years.
(e) Not enough money is being spent on research and development of alternative fuels.
(f) It is possible to increase energy supplies while protecting the environment at the same time.
(g) New technologies will make it possible to have enough electricity for all of us in the future.

Respondents were given the following response alternatives: 1 = Strongly disagree, 2 = Somewhat disagree, 3 = Neutral, 4 = Somewhat Agree, 5 = Strongly Agree. Table 10.1 shows the average response for each of the four state samples in our survey, with possibilities ranging from 1 to 5.

California had statistically significant energy policy beliefs when compared to Washington on the question, "I am concerned about being personally affected by shortage of electricity in the next five years," with the California respondents more concerned than those in Washington. California respondents also were more likely than those in Idaho to agree that "Not enough money is being spent on research and

Table 10.1 Average responses of state publics to energy policy questions

Policy Question	California	Oregon	Washington	Idaho	Sig.
(a) National security[a]	4.44	4.40	4.46	4.36	.159
(b) Foreign ownership[a]	4.38	4.36	4.38	4.38	.944
(c) U.S. lacks resources[a]	3.50	3.53	3.46	3.42	.447
(d) Electricity shortage[b]	3.19	3.06	2.95	3.05	.003
(e) Fund alt. energy[c]	3.77	3.75	3.62	3.59	.005
(f) Increase energy and protect environment[a]	4.27	4.16	4.24	4.18	.084
(g) Technology will provide needed energy[c]	4.06	3.93	3.93	3.89	.010
N=	686–690	752–754	710–714	685–687	

[a]No two state differences are statistically significant
[b]California and Washington are different at the .05 level (Scheffe)
[c]California and Idaho are different at the .05 level (Scheffe)

Table 10.2 Energy policy factor loadings of varimax rotation of principal components factor analysis

	Factor		
	I	II	III
	Scarcity	Security	Technology
Decreasing our dependence on foreign oil and gas is important to our national Security.	.08	**.86**	.09
I am concerned about foreign ownership of our energy resources.	.17	**.82**	.13
I am concerned that our country doesn't have enough energy resources.	**.79**	.16	−.06
I am concerned about being personally affected by shortage of electricity in the next 5 years.	**.76**	.09	.00
Not enough money is being spent on research and development of alternative fuels.	**.62**	.04	.35
It is possible to increase energy supplies while protecting the environment at the same time.	−.07	.14	**.81**
New technologies will make it possible to have enough electricity for all of us in the future.	.17	.07	**.77**
Eigenvalues	2.23	1.24	1.05
% of Variance	31.8%	17.8%	14.9%

N = 2832–2846

development of alternative fuels" and that "New technologies will make it possible to have enough electricity for all of us in the future." Overall, though, the absolute differences among the states were relatively small.

Rather than continuing to analyze the individual items, factor analysis was used to explore the responses of the total four-state sample to the seven questions to assess whether there was an underlying structure or common content to all or sets of responses. We used principal components factor analysis with a varimax rotation to maximize the clarity of the patterns. The results are shown in Table 10.2.

Three dimensions emerge from the factor analysis shown in Table 10.2. The most central dimension is labeled a *scarcity* factor, reflecting perceptions of the country as not having enough resources, concern about being personally affected by a shortage of electricity, and not enough money being spent on alternative fuel research. The second dimension reflects concerns about *national security* in the sense of dependence on foreign oil and gas and that decreasing that dependence is important to our national security. The third dimension is centered on *views of technology*, including the belief that energy supplies can be increased, even while protecting the environment, and that new technologies will make it possible to have enough electricity in the future. The question for this chapter is whether variations of individuals on those three dimensions enhance the capacity to explain their views on the tradeoffs involved in AETs. In particular, one would hypothesize that greater support for AETs, even in the face of potential near-future environmental damage, would more likely occur in those respondents seeing likely greater scarcity of

Table 10.3 Correlations of public's energy policy beliefs with political, NEP, postmaterial and climate change orientations

	Energy policy dimension		
	Scarcity	Security	Technology
Environmental orientation (NEP)	.31**	.02	.09**
Postmaterial values	.00	.07**	.05**
Political orientation (lib/con)	−.18**	.02	−.14**
Climate change perception	−.24**	−.01	−.07**

**p ≤ .01; N = 2790–2829

Table 10.4 Correlation of background attributes with positions of the public on energy policy dimensions

	Attributes					
	Age	Gender	Education	Residence	Employed	Income
Scarcity	−.04*	−.18**	−.10**	−.03	.00	−.10**
Security	.06**	.00	−.01	.05**	.03	.04
Technology	−.04*	−.04*	−.04*	.01	−.02	−.01

* p ≤ .05; **p ≤ .01; N = 2688–2727

energy, greater danger from foreign control of energy, and greater capacity of technology to solve the problems.

Are these energy policy perceptions connected to the respondents' background attributes and to their more general orientations and beliefs? Thus, for example, do those respondents with higher levels of educational achievement hold different beliefs about the capacity of technology to provide sufficient electricity for everyone for the future? Or, are individuals with more conservative political orientations more likely to perceive a threat to national security from energy resource scarcity or from dependence on foreign sources of oil and gas? And, do individuals with pro-environmental values on the NEP hold different views of the energy policy arena? The results shown in Tables 10.3 and 10.4 allow answers to those and other questions. The factor scores associated with the three dimensions are used to analyze energy policy beliefs.

Each of the general political and social orientations was significantly correlated with two of the three energy belief dimensions. The NEP environmental orientation was strongly associated (r = .31) with the scarcity dimension—environmentalists were more likely to see a scarcity of resources—and less so, but still at a significant level, with the technology dimension (r = .09)—environmentalists had more faith in technology. Postmaterial values were weakly associated with the security and technology dimensions. Liberals were more likely than conservatives to see scarcity of energy as a threat, and less likely to see technology as a possible solution. And, those respondents who agreed that the earth is getting warmer after the past few decades were more likely to see scarcity in energy resources, and slightly more likely to see technology as a potential solution to that problem. Thus, the public's

Table 10.5 Correlations (r) among AET tradeoff opinions and energy beliefs factor scores: scarcity, security, and technology

	Energy policy beliefs		
AET tradeoffs	Scarcity	Security	Technology
Index	.04*	−.02	.02
Wind	−.05*	.00	−.06**
Solar	.03	.01	−.03
Hydro	.03	.01	−.03
Wave	.00	.00	−.08

* $p = \leq .05$; ** $p \leq .01$; N = 2734–2762

views of energy policy seem at least partially rooted in their more general political orientations.

As shown in Table 10.4, age was the only background attribute with a statistically significant relationship with each of the three energy policy dimensions, but none of them was particularly powerful. Older individuals were a little less likely to see scarcity as an issue and to see technology as a solution, but were more likely to see national security as a threat from energy shortages. Women were more likely than men to see scarcity as an important possibility; those with more education were less likely to see scarcity as an issue, and those with greater income levels were less likely to see scarcity as a problem. Overall, then, there were some gulfs among individuals with different background attributes, but those gaps were not particularly large.

In terms of the central question considered here, though, do these energy policy beliefs make any difference in how members of the public responded to the tradeoffs between long-term environmentally driven AETs and the projected short-term negative environmental effects of those technologies? To answer this question, we first correlated the tradeoff responses with the three energy policy factor scores for scarcity, security, and technology. The results are shown in Table 10.5.

While there were several statistically significant relationships in Table 10.5, none were very strong, and the directions of those relationships were mixed. Thus, the overall AET tradeoff index had a weak positive relationship to scarcity, but the wind responses were weakly associated in a *negative direction* with both scarcity and technology. Respondents who saw energy scarcity as a bigger likelihood were marginally more likely to support the development of a "high potential wind energy farm in windy, rare tall grass prairie area." Likewise, the same was true for respondents who saw technology as a potential route to energy production. The analysis further explored whether those relationships were truly linear—that is, whether an increase in the support for the wind installation was associated with an increase in the perception of energy scarcity across the full range of that support. The results are shown in Table 10.6, which presents the mean scores for the three energy policy dimensions within each category of response to the tradeoff questions.

The various alternative energy technology tradeoff questions produced quite different results. Thus, the relative perception of impending scarcity was unrelated to

Table 10.6 Mean energy policy scores within responses to AET tradeoffs

	A. Wind farms in Tall Grass Prairie					
	1	2	3	4	5	
	Develop wind farms				Protect tall grass prairie	F
Scarcity	.06	.02	−.01	−.04	−.11	1.4
Security	.12	−.07	−.06	.07	−.04	3.8**
Technology	.26	−.00	.07	−.16	.19	13.5**
	B. Solar power farm even if disrupts endangered turtles					
	1	2	3	4	5	
	Install solar farm				Protect turtles	F
Scarcity	−.13	.01	−.03	.21	−.21	9.0***
Security	.12	−.03	−.05	−.08	.27	6.8***
Technology	.19	.05	−.14	.02	.16	10.1***
	C. Off-shore wave turbines even if impedes whales					
	1	2	3	4	5	
	Install turbines even if disturbs Whales				Protect whales	F
Scarcity	−.08	−.02	−.01	.08	−.07	2.2
Security	.28	.04	−.09	.02	.05	4.8***
Technology	−.11	−.08	−.08	.07	.12	5.7***
	D. Hydroelectric even if harms salmon					
	1	2	3	4	5	
	Continue hydroelectric even if harms salmon				Remove dams	F
Scarcity	−.44	.03	.06	.05	.05	20.5***
Security	.26	−.01	−.12	.05	.01	9.9***
Technology	−.02	−.03	−.08	.11	.13	4.2**

N = 2734–2762

preferences as to whether wind farms should be built, even if in native tall grass prairie. On the other hand, national security appeared to be a minor driver of the belief that wind farms should be installed, even if they disturb the tall grass prairie. The most interesting pattern, though, had to do with technology. Support for the installation of wind farms was associated with both ends of the technology dimension. When compared to the respondents in the middle of the wind tradeoff scale, belief in the capacity of technology to help solve possible energy shortages was higher both among those who strongly supported the installation of wind farms and among those who strongly opposed the installation of wind farms when they disturb the tall grass prairies.

The energy belief patterns among various positions on the tradeoff of solar installations versus their impact on endangered turtle habitat also had some unusual characteristics. In particular, regarding the scarcity dimension, the two extreme positions on solar energy were similar in their lesser concern that there is or will be a shortage of energy resources. Those respondents at the extreme end of protecting turtles exhibited the lowest concern with energy scarcity. The same curvilinear relationship was obtained for both the security and the technology dimensions: the extreme positions on solar tradeoffs were more likely to both express concerns about foreign ownership of resources and to believe that technology can be a solution to energy shortages than were those near the middle.

The wave energy question showed that where the respondent stood on the tradeoff with whale migration had no impact on whether energy scarcity was a personal concern. On the other hand, those who believed that the choice should be to "install off-shore wave turbines on the sea floor to produce electricity even if it impedes whale migration" scored much higher on the energy-based national security factor dimension. On the technology factor score, those at the "install" end were significantly lower than were those at the end that saw technology as a potential solution to energy shortages.

Perhaps the most significant finding was obtained in the different positions on the use of hydroelectric dams, even if it harms salmon runs. Those who were at the lowest position ("continue to operate hydroelectric dams for power even if it harms salmon runs") had a far smaller mean (−.44) on the scarcity dimension than did those at any other position.

Further analysis shows that the New Ecological Paradigm continues to have a powerful effect on AET tradeoffs, even when considering the respondents' energy policy beliefs. At the same time, however, those energy policy beliefs have tradeoff effects as well, sometimes mitigating or altering the impact of the NEP for certain technologies. Table 10.7 shows those effects in the form of a series of analyses of variance in which the dependent variable is an average score on the tradeoff questions calculated for each combination of positions (low, medium, high) on the NEP and the Energy Policy dimensions (scarcity, security, and technology).

10.3.1 Scarcity

Summary Index Across the four alternative energy technologies summative index, the perception of scarcity had a very different affect at the high level of the New Ecological Paradigm than it had at the low and middle levels. At the high level of the scarcity dimension, the higher NEP respondents exhibited greater support for the environmental preservation end of the index. However, at the lower and middle levels of scarcity, the high NEP respondents were *less* likely to support the preservationist position. Thus, there is a strong and somewhat counterintuitive interaction effect between scarcity and environmental values (Adjusted $R^2 = .18$; Significance: NEP = .000; Scarcity = .297; Interaction = .000). This unexpected pattern may in fact result from contrasting results for the specific AETs and/or the specific short-term negative environmental effects they are seen as producing.

Wind Farms Among low NEP respondents, high scarcity increased support for wind installation even if it threatened tall grass prairie ecology, but among high

Table 10.7 Analysis of variance results of the effect of the new ecological paradigm and energy policy beliefs on alternative energy technology tradeoffs

Wind	Scarcity			
	NEP	Low	Medium	High
	Low	2.68	2.39	2.28
	Medium	2.83	2.84	2.56
	High	2.91	3.17	3.15
	R^2 = .06; F, NEP = 74.3, p = .000; F, Scarcity = 4.23, p = .020; F, Interaction = 9.13, p = .000			
	Security			
	NEP	Low	Medium	High
	Low	2.62	2.36	2.52
	Medium	2.72	2.78	2.68
	High	3.05	3.11	3.11
	R^2 = .04; F,NEP = 56.97, p = .000; F, Security = .719; p = .487; F, Interaction = 2.08, p = .080			
	Technology			
	NEP	Low	Medium	High
	Low	2.68	2.45	2.34
	Medium	2.81	2.67	2.76
	High	3.18	3.11	3.04
	R^2 = .05; F,NEP = 62.63, p = .000; F, Technology = 6.16, p = .002; F, Interaction = 1.52, p = .19			
Solar	Scarcity			
	NEP	Low	Medium	High
	Low	2.48	2.57	2.41
	Medium	2.92	2.81	2.78
	High	3.21	3.26	3.29
	R^2 = .08; F,NEP = 101.93, p = 000; F, Scarcity = .63, p = .53; F, Interaction = 1.44, p = .217			
	Security			
	NEP	Low	Medium	High
	Low	2.56	2.58	2.31
	Medium	2.85	2.80	2.82
	High	3.36	3.24	3.19
	R^2 = .08; F, NEP = 113.03, p = .000; F, Security = 4.88, p = .008; F, Interaction = 1.79, p = .128			
	Technology			
	NEP	Low	Medium	High
	Low	2.66	2.56	2.15
	Medium	3.00	2.69	2.81
	High	3.11	3.10	3.46
	R^2 = .10; F, NEP = 111.8, p = .000; F, Technology = 4.58, p = .010; F, Interaction = 16.21, p = .000			

(continued)

Table 10.7 (continued)

Wave	Scarcity			
	NEP	Low	Medium	High
	Low	3.12	3.06	2.95
	Medium	3.68	3.41	3.28
	High	3.73	3.80	3.82
	R^2 = .08; F, NEP = 105.21, p = .000; F, Scarcity = 4.85, p = .008; F, Interaction = 4.06, p = .003			
	Security			
	NEP	Low	Medium	High
	Low	3.00	3.22	3.00
	Medium	3.41	3.49	3.43
	High	3.81	3.75	3.82
	R^2 = .07; F, NEP = 97.55, p = .000; F, Security = 1.54, p = .216; F, Interaction = 1.78, p = .131			
	Technology			
	NEP	Low	Medium	High
	Low	2.97	3.03	3.25
	Medium	3.33	3.39	3.62
	High	3.77	3.74	3.84
	R^2 = .08; F, NEP = 90.61, p = .000; F, Technology = 11.17, p = .000; F, Interaction = .991, p = .411			
Hydro	Scarcity			
	NEP	Low	Medium	High
	Low	2.31	2.36	2.63
	Medium	2.93	2.99	2.80
	High	3.14	3.45	3.53
	R^2 = .13; F, NEP = 148.87, p = .000; F, Scarcity = 7.29, p = .001; F, Interaction = 6.42, p = .000			
	Security			
	NEP	Low	Medium	High
	Low	2.38	2.52	2.29
	Medium	2.98	2.85	2.87
	High	3.65	3.38	3.24
	R^2 = .13; F, NEP = 188.78, p = .000; F, Security = 7.90, p = .000; F, Interaction = 3.62, p = .008			
	Technology			
	NEP	Low	Medium	High
	Low	2.32	2.38	2.52
	Medium	2.84	2.82	3.05
	High	3.47	3.38	3.40
	R^2 = .13; F, NEP = 180.33, p = .000; F, Technology = 3.65, p = .026; F, Interaction = 1.85, p = .117			

(continued)

Table 10.7 (continued)

Tradeoff Summary Index	Scarcity			
	NEP	Low	Medium	High
	Low	10.57	10.38	10.30
	Medium	12.76	12.06	11.41
	High	12.97	13.68	13.78
	R^2 = .18; F, NEP = 237.97, p = .000; F, Scarcity = 1.22, p = .297; F, Interaction = 7.63; p = .000			
	Security			
	NEP	Low	Medium	High
	Low	10.57	10.70	10.10
	Medium	12.02	11.92	11.81
	High	13.87	13.46	13.36
	R^2 = .17; F, NEP = 261.58, p = .000; F, Security = 4.86, p = .018; F, Interaction = 1.14, p = .335			
	Technology			
	NEP	Low	Medium	High
	Low	10.62	10.41	10.29
	Medium	11.99	11.56	12.26
	High	13.52	13.33	13.72
	R^2 = .17; F, NEP= 255.6, p= .000; F, Technology= 3.58, p= .028; F, Interaction= 1.86, p = .115			

NEP respondents, high scarcity increased support for preserving that prairie and not installing the wind farms (Adjusted R^2 = .06; Significance: NEP = .000; Scarcity = .020; Interaction = .000). Thus, the level of affirmation of NEP values structured the impact of perceived energy scarcity on the position one takes on the tradeoffs involved in wind installations. Scarcity failed to overcome the reluctance of high NEP respondents to approve wind farms when they had putative short-term negative environmental consequences, even when those wind farms were argued to pose much greater environmental returns over the long run.

Solar In contrast to the case with wind farms, the relative perception of energy scarcity had no significant effect on the role of the NEP on relative support for positions on the solar tradeoff question (Adjusted R^2 = .08; Significance: NEP = .000; Scarcity = .533; Interaction = .217). So, one might wonder why the role of scarcity was significantly less important in mitigating NEP's effects when it came to solar than when it concerned wind farms? Perhaps the answer is found in the tradeoff at issue. Is there something about "tall grass prairie" that elicits a different response than "the ecology of endangered turtles?" That difference may be the seemingly inanimate nature of a tall grass prairie at risk when compared to the "lives" of turtles. Perhaps the perceived scarcity of energy supplies is more important to high NEP people than the tall grass prairie, but not more important than the endangered turtle.

Wave The perception of likely energy scarcity affected the low and mid-level NEP adherents, decreasing in both the support for protecting whale migration at the expense of the installation of wave turbines. On the other hand, in the high NEP category of respondents, high scarcity showed a slightly greater propensity to protect the whales (Adjusted R^2 = .08; Significance: NEP = .000; Scarcity = .008; Interaction = .003). Clearly, the biggest change was obtained among those at the middle level of the NEP, where high scarcity respondents were much less likely than the low scarcity ones to sacrifice whale migratory routes in the pursuit of alternative energy that presumably would help satisfy the perceived energy shortfalls.

Hydro Unlike the patterns for wind farms and wave turbines, among both the high NEP and the low NEP groups, the perception of energy scarcity substantially increased support for removal of dams where they cause harm to salmon runs (Adjusted R^2 = .13; Significance: NEP = .000; Scarcity = .001; Interaction = .000). This pattern may stem both from the pervasive presence of hydroelectric dams across the region and from the unique and central historic role of salmon in the culture of the West Coast states. Indeed, the concept of scarcity applies both to the supply of energy and to the presence of the anadromous fisheries that filled the rivers with outlets to the Pacific Ocean. Indeed, it is very likely that energy produced from hydroelectric dams that disrupt anadromous fisheries is not seen as "an alternative energy technology" in the West. While not related to anadromous fisheries, the following excerpt from a 2012 story in the *Aspen Times* reflects the division in environmentalists' views of hydropower (Condon 2012):

> One faction understands the size and scope of climate change, he said. The other remains focused on land conservation issues – creating wilderness and protecting streams. Those focused on climate change are more willing to see wind farms off the Massachusetts coast and giant solar farms in the California desert as worthwhile steps in the battle against climate change…

On the other hand, the National Hydropower Association (2014) reports that "support for hydropower is slightly stronger in the Northwest—WA, OR, ID" and that "75% of registered voters say that an "excellent reason to favor the use of hydropower" is because it is a "clean" energy source – 64% cite "efficient" as an excellent reason and 62% because it is "renewable." Moreover, "support for hydropower as an energy source grows by 5% (89–94%) after registered voters are informed about all of hydropower's benefits AND impacts on the environment."

10.3.2 Security

Summary Index At each level of the NEP index, a greater belief that the country's security is affected by potential energy shortages produced less support for the preservation position on the overall AET tradeoff index. (Adjusted R^2 = .17; F, NEP = 261.58, p = .000; F, Security = 4.86, p = .008; F, Interaction = 1.14, p = .335). Those differences were substantially greater at the low and high ends of the NEP, compared to the middle third. Thus, across the board, higher concern with energy-based security suppressed the shorter-term impact of the NEP. Even so, at each level of security concern, the NEP level had a much stronger impact than the security beliefs on overall tradeoff levels.

Wind Security beliefs had no significant effect on the impact of the NEP on tradeoff beliefs in regard to the installation of wind technology (Adjusted R^2 = .04; F, NEP = 56.97; p = .000; F, Security = .719, p = .487; F, Interaction = 2.08, p = 080). At each level of the security dimension, the High NEP respondents were more likely to take the preservationist position. On the other hand, at each level of the NEP, there was little difference between the levels of the security dimension in their position on the wind tradeoff question.

Solar Contrary to the case with regard to wind installations, security beliefs did have a significant effect on the impact of the NEP on tradeoff beliefs in regard to solar technology (Adjusted R^2 = .08; F, NEP = 101.93, p = .000; F, Security = 4.88, p = 008; F, Interaction = 1.79, p = 128). At the two end categories (high and low) of the NEP, the increase in security beliefs diminished the concern for the short-term environmental tradeoffs associated with the installation of solar energy technology. To be sure, however, once again the NEP values were much more significant in their influence on solar tradeoffs than were the security beliefs.

Wave Security beliefs had no significant effect on the impact of the NEP on tradeoff beliefs in regard to the installation of wave turbine technology (Adjusted R^2 = .07; F, NEP = 97.55, p = .000; F, Security = 1.54, p = .216; F, Interaction = 1.78, p = .131). To be sure, the effect of the NEP itself is substantial. While respondents were not queried as to the specific security implications of the wave technology, the relative newness of the proposed technology surely would result in little more than the most superficial understanding.

Hydro Security beliefs had small negative effects on support for removing dams on the low and middle levels of NEP orientations, but a very large effect in the highest level of support for the New Ecological Paradigm (Adjusted R^2 = . 13; F, NEP = 188.33, p = .000; F, Security = 7.9, p = .000; F, Interaction = 3.62, p = .008). Even so, the high security concern/high NEP group expressed higher support for removing the dams than any other group, save for the low and medium security respondents at the high NEP level. And, the lowest support for the removal of the dams was in the low NEP/high security corner, in contrast to the opposite corner of low security/high NEP, which expressed the highest support for dam removal.

10.3.3 Technology

Summary Index On the overall summary index, the impact of pro-technology beliefs had a different effect at different levels of the New Ecological Paradigm. Thus, at the lowest NEP category, the higher technology belief category showed somewhat *lower* preservationist positions, while in the medium and high NEP levels, those same technology beliefs were associated with somewhat greater preservationist views (Adjusted $R^2 = .17$; F, NEP = 255.6, p = .000; F, Technology = 3.58, p = .028; F, Interaction = 1.86, p = .115). How to explain those differences? From a particular perspective, it makes sense that respondents in the low NEP category would have less concern about the putative negative short-term environmental effects of AETs precisely because of NEP locations. On the other hand, those at the highest NEP levels along with the highest technology level may actually believe that technology would make it possible to solve energy problems without encumbering negative effects on the environment. Therefore, if those negative effects are identified, they would reject the use of that particular technology.

Wind Belief that technology will be an instrument for dealing with possible future energy shortages seemed to have its greatest effect at the low level of NEP beliefs, where those in the high *technology* category were less likely to take the preservation position on wind installations. The middle and high levels of the NEP exhibited similar effects, but those were much less strong. (Adjusted $R^2 = .05$; F, NEP = 62.63, p = .000; F, Technology = 6.16, p = .002; F, Interaction = 1.5, p = .19). There was an apparent logic to this pattern, in that at the low NEP level, the values would be much less likely to interfere with the link between technology beliefs and orientations in regard to the installation of that technology, independent of the more short-term effects on tall grass prairie.

Solar Solar energy technology beliefs had quite contrasting effects at different levels of support for the New Ecological Paradigm. Again, just as in the case of the wind installations, at the lowest NEP level, individuals in the high category on technology were much *less* likely to take the preservationist position on solar technology installation. And, on the other hand, at the high NEP level, the high technology respondents were *more* likely to take the preservationist position (Adjusted $R^2 = .10$; F, NEP = 111.8, p = .000; F, Technology = 4.58, p = .010; F, Interaction = 16.21, p = .000). Overall, the effects in regard to technology beliefs were substantially stronger in the case of solar than in the case of wind farms. Perhaps that difference has to do with the particular environmental tradeoffs in the two cases—wind farms being the prairie and solar being the turtles.

Wave In contrast to the patterns with wind farms and solar installations, at the low and mid-levels of the NEP, technology beliefs were associated with respondents being *more* likely to take the preservationist position on wave energy (Adjusted $R^2 = .08$; F, NEP = 90.61, p = .000; Technology = .000; Interaction = .411). Again, we are pressed to try to answer the question of whether it is the nature of the tradeoff (whales versus prairie or turtles), the nature of the environment in which they reside or are located (water versus land), or the nature of the technology. In the latter case, perhaps those pro-technology respondents are more tuned into the possibility of negative effects from the installation of wave turbines.

Hydro As in wave alternative energy, the impact of technology beliefs was similar in the two lower levels in being associated with a somewhat greater likelihood of preservationist beliefs (Adjusted $R^2 = .13$; F, NEP = 180.33, p = .000; F, Technology = 3.65, p = .026; F, Interaction = 1.85, p = .117). On the other hand, very minimal differences in hydro tradeoff beliefs were obtained among the technology levels within the high NEP values category.

The final research question to be addressed here returns us to the differences between: (1) the *globalists* (those who express strong environmental values as measured by the NEP *and* support the installation of AETs, even if they cause localized environmental costs); and (2) the *localists* (those who express strong environmental, values as measured by the NEP, but reject AETs if they impose local environmental costs). Table 10.8 shows the average scores of the two groups on the three variables of concern here: energy scarcity, energy security, and the power of technology to solve energy shortages. Recall that those scores were derived from a factor analysis of multiple items.

The only globalist/localist difference of any magnitude was obtained in their beliefs in regard to energy scarcity, with the localists more likely to see looming energy scarcity than the globalists (p = .09). At first blush, that may seem an anomalous finding. Why would those individuals who exhibited weaker support for AETs in the face of localized environmental costs be more concerned about energy scarcity than those who expressed greater support for AETs designed to mitigate energy shortages produced by the exhaustion of traditional sources or by limitations? One possibility is that globalists are less likely than localists to see climate change (72–88%), and they are less likely to see climate change as caused by human activity (64% compared to 82%). So, localists may also believe, to a greater degree, that climate change can be halted through individual behavior, rather than by local—environment damaging—technological solutions, such as those central to alternative energy installations.

Table 10.8 Average energy belief scores among globalists and localists[a]

Energy beliefs	Globalists	Localists	Sig.
Scarcity	.14	.30	.09
Security	.14	.11	.82
Technology	.08	.10	.85

[a]The entries are factor scores produced from a varimax rotation of factor analysis of seven beliefs about current energy conditions and policy.

10.4 Conclusion

This chapter examined the impact of energy policy beliefs on individual preferences regarding alternative energy technologies within the context of more general support for environmental values. The energy policy beliefs in question concerned the likely scarcity of energy resources, the concern about likely dependence on foreign sources of energy, and the potential for technology to provide solutions to the need for significant energy resources. Compared to security and technology beliefs, scarcity had the strongest set of correlations with environmental orientations (New Ecological Paradigm), political orientations (as measured by liberal/conservative), and perceptions of climate change. Overall, the correlates of energy policy beliefs and AET beliefs were rather weak, although some individual patterns in the data proved to be important. When NEP values and AET attitudes were combined into globalist and localist types, only the energy policy belief of scarcity distinguished between the two groups, with localists perceiving greater likelihood of insufficient resources.

References

Ansolabehere, S., & Konisky, D. M. (2014). *Cheap and clean: How Americans think about energy in the age of global warming*. Cambridge, MA: The MIT Press.

Center for American Progress. (2014). Public opinion on U.S. energy and environmental policy: Key findings from a nationwide survey among 1,101 likely 2016 voters, conducted December 5 to 9, 2014. http://www.climateaccess.org/sites/default/files/Hart_Public%20opinion%20 US%20energy%20policy.pdf. Accessed 15 Aug 2016.

Condon, S. (2012, November 2). Aspen hydro plant splits environmentalists. *The Aspen Times*. www.aspentimes.com/article/20121102/NEWS/1212209980. Accessed 15 Aug 2016.

Council on Foreign Relations. (2009). U.S. opinion on energy security. In*Public opinion on global issues*. New York: Council on Foreign Relations.

Dryzek, J. S. (2002). A post-positivist policy-analytic travelogue. *The Good Society, 11*, 32–36.

Frewer, L. J., Howard, C., & Shepard, R. (1998). Understanding public attitudes to technology. *Journal of Risk Research, 1*, 221–235.

Gallup Poll. (2015). Americans choose environment over energy development. http://www.gallup.com/poll/182402/americans-choose-environment-energy-development.aspx. Accessed 15 Aug 2016.

Gallup Poll. (2016). Energy. http://www.gallup.com/poll/2167/energy.aspx. Accessed 15 Aug 2016.

Johansson, B. (2013a). A broadened typology on energy and security. *Energy, 53*, 199–205.

Johansson, B. (2013b). Security aspects of future renewable energy systems—A short overview. *Energy, 61*, 598–605.

National Hydropower Association. (2014). *About hydro power*. www.nahydro.com/public-info/hydropower-facts. Accessed 15 Aug 2016

Nisbett, M. C., Maibach, E., & Leiserowitz, A. (2011). Framing peak petroleum as a public health problem: Audience research and participatory engagement in the United States. *American Journal of Public Health, 101*, 1620–1626.

On the Issues. (2010). Sara Palin on Energy and Oil. 2010 Tea Party Convention. http://www.ontheissues.org/Celeb/Sarah_Palin_Energy_+_Oil.htm. Accessed 15 Aug 2016.

Perry, M. (2013, November 19). From energy scarcity to a new era of energy abundance: 'Saudi America' should embrace its energy superpower status. *AEI Ideas*. http://www.aei.org/publication/from-energy-scarcity-to-a-new-era-of-energy-abundance-saudi-america-should-embrace-its-energy-superpower-status/. Accessed 15 Aug 2016.

Pew Research Center. (2015, January 15). *How Americans view the top energy and environmental issues*. http://www.pewresearch.org/topics/energy-and-environment/. Accessed 15 Aug 2016.

Porwitt, J. (2000). *Playing safe: Science and the environment*. New York: Thames and Hudson.

Steel, B. S., Lach, D., & Satyal, V. (2006). Ideology and scientific credibility: Environmental policy in the American pacific northwest. *Public Understanding of Science, 15*, 481–495.

Steel, B. S., Pierce, J. C., Warner, R. L., & Lovrich, N. P. (2015). Environmental value considerations in public attitudes about alternative energy development in Oregon and Washington. *Environmental Management, 55*, 634–645.

Texas A&M University. (2012). Texas A&M poll shows Americans support renewable energy—But don't want to pay for it at the pump. *Texas A&M University National Energy Opinion Poll*. http://engineering.tamu.edu/news/2012/11/09/texas-am-poll-shows-americans-support-renewable-energy-—-but-dont-want-to-pay-for-it-at-the-pump. Accessed 15 Aug 2016.

University of Texas. (2016). Majority of Republicans and Democrats say government and industry should work together to improve energy security. University of Texas Austin. Energy Poll. http://news.utexas.edu/2016/02/18/ut-energy-poll-reveals-opinions-on-energy-security. Accessed 15 Aug 2016.

Whitehouse. (2011) Remarks by the President on America's Energy Security. Georgetown University. https://www.whitehouse.gov/the-press-office/2011/03/30/remarks-president-americas-energy-security. Accessed 15 Aug 2016.

Chapter 11
Knowledge: Levels and Sources

Abstract Energy policy questions, including the siting of alternative energy installations, often are complex as well as scientific and technical in nature. At the same time, significant ideological and political pressures produce claims on behalf of the rights of citizens to be involved in decisions about such policies and their implementation. Elsewhere, this has been labeled as the "democracy and the technical information quandary," namely, how does one legitimately enhance the involvement of the general public in a policy area in which many of its members surely hold a knowledge deficit? But further questions are stimulated as well. Thus, does knowledge of a policy area make any difference in the opinions individuals express? Does knowledge level make any difference in the degree to which individuals can link their policy preferences to their fundamental values? To answer these and other questions, this chapter begins with a look at the knowledge and familiarity citizens bring to their attitudes toward energy policy and energy conditions. This chapter also examines whether knowledge levels are related to respondents' policy positions on the alternative energy tradeoff preferences.

Keywords Democracy and technical information quandary • Energy policy • Policy relevant knowledge

11.1 Introduction

> *We're loaded and we didn't even know it. (2016 Presidential candidate Donald Trump commenting on fracking* (Egan 2016))

> *To ask people their views about energy without asking about their views regarding climate change is more than just a careless oversight. It misses what is perhaps now the central issue in energy: Can we as a civilization survive the long-term side-effects of the fuels we currently use?* (Columnist Kurt Cobb 2016)

In *What Americans Know About Politics and Why it Matters*, Delli Carpini and Keeter argue that public knowledge of important political and policy issues "is essential for citizens to discern their real interests," as well as to effectively involve themselves in the policy process (1996, p. 3). Given that most scientists argue that human caused greenhouse gas emissions are the main source of climate change, it

J.C. Pierce, B.S. Steel, *Prospects for Alternative Energy Development in the U.S. West*, Environmental Challenges and Solutions 4,
DOI 10.1007/978-3-319-53414-5_11

is important to assess the scope and depth of policy-relevant energy knowledge among the public, and to identify the link between policy-relevant knowledge and support for renewable energy technologies. By more clearly specifying the connection between policy relevant knowledge levels and the impact—if any—on beliefs about alternative energy technology (AET) tradeoffs, one can better understand how information dissemination efforts may be designed to more effectively engage the public in AET siting issues and to assist the public in understanding policy discussions that concern AET.

Energy policy questions, including the siting of alternative energy installations, often are complex as well as scientific and technical in nature. At the same time, significant ideological and political pressures produce claims on behalf of the rights of citizens to be involved in decisions about such policies and their implementation. Elsewhere, this has been labeled as the "democracy and the technical information quandary" (Pierce and Lovrich 1986), namely, how does one legitimately enhance the involvement of the general public in a policy area in which many of its members surely hold a knowledge deficit? But further questions are stimulated as well. For example, does knowledge of a policy area make any difference in the opinions individuals express? Does knowledge level make any difference in the degree to which individuals link their policy preferences to their fundamental values? To answer these and other questions, this chapter begins with a look at the knowledge and familiarity citizens bring to their attitudes toward energy policy and energy conditions. We will also examine whether knowledge levels are related to respondents' policy positions on AET tradeoff preferences.

11.2 Prior Research

As discussed in previous chapters in this book, most research on public attitudes and knowledge of energy policy has focused on traditional sources of energy: fossil fuels such as petroleum, natural gas, and coal (e.g., Smith 2002). When examining the literature on public attitudes toward alternative energy sources, Bolson and Cook found that "with the exception of attitudes toward nuclear energy, very little has been asked by polling organizations about alternative energy sources" (2008, p. 373). According to Ansolabehere and Konisky, another problematic issue with much of the survey research is "...the lack of head-to-head comparisons of energy sources (2014, p. 49). The recent public dialogue concerning climate change has generated some rudimentary polling results concerning energy conservation and the development of renewable sources by the Pew Research Center for the People and the Press (e.g., 2008). But these polls have not examined the effects of knowledge and values.

In many policy subsystems—such as energy policy and the development of alternative energy technologies—the scientific and technical complexity of issues poses serious challenges for the effective participation of citizens in the democratic process. While in a democracy citizens have the right to engage in the policy process and influence policy-makers, the public can be at a disadvantage if its members are ill informed on policy options and the scientific and technological aspects of those options. The gap between the public's need for policy-relevant information and its-less-than-optimal levels of policy-relevant knowledge has led some social scientists to proclaim the existence of a "legitimacy crisis" (Dahl 1985). In fact, Janicke has argued: "Without knowledge there is no (perceived) problem, no public awareness, and consequently no policy process..." (1997, p. 7). This line of argumentation reflects the assumption that policy-relevant knowledge is essential for citizens' understanding of policy alternatives and thus to the public's ability to help frame and influence the outcome of policy processes.

While policy-relevant knowledge has been shown to enhance the ability of citizens to recognize and act on their values and interests, much research on the distribution of public knowledge levels in a number of policy areas has documented a "knowledge gap" separating persons of lower and higher socioeconomic status (Delli Carpini and Keeter 1996; Pierce et al. 1989, 1992). Persons of lower socioeconomic status (SES) have typically been shown to exhibit lower levels of knowledge holding, and they also tend to respond more slowly than persons of higher SES to efforts to increase the public's knowledge (Genova and Greenberg 1979). In fact, many variables associated with SES—level of educational attainment, income, and occupational status—have been found to be associated with environmental awareness in the United States, Canada, Japan, and Russia (Delli Carpini and Keeter 1996; Pierce et al. 1992, 2000). Gender and age also have been found to be important correlates of policy-relevant knowledge holding, with young cohorts exhibiting lower levels of policy relevant knowledge, compared to older cohorts, and men tending to have higher levels of knowledge than women about political process and public policy issues (Delli Carpini and Keeter 1996; Jamieson 2000).

If policy-relevant knowledge concerning renewable energy policy issues is deeply rooted in SES and demographic attributes alone, the prospects for increasing public knowledge levels will be limited, due to the static character of gender, age, and educational attainment for individuals. In addition, if socioeconomic classes do indeed differ markedly in their respective knowledge levels, individuals in the knowledge-deficient classes may be less able to protect and promote their interests than better-informed classes (Delli Carpini and Keeter 1996). The less familiar the policy alternatives, their underlying assumptions, and their implications for quality of life and community economic fortunes, the greater will be the difficulty for the information-disadvantaged to overcome the knowledge deficit.

In addition to SES and demographic explanations for policy relevant knowledge, some social scientists have articulated another explanation for knowledge levels that includes values and ideology consideration, such as the New Ecological Paradigm and political orientations (Ettema and Kline 1977; Pierce et al. 2000; Steel et al. 2006). This research suggests that even when controlling for SES characteristics and demographics, value and ideological variation can exhibit independent effects (Steel et al. 2006). Value attributes can lead to the acquisition of information by citizens irrespective of their SES and demographic characteristics. For example, knowledge-seeking and knowledge-holding is highest among individuals who see a particular stake in policy outcomes (Steel et al. 1990) and those who are strongly committed to their policy views (Pierce et al. 1992). Thus, one might expect higher levels of knowledge concerning renewable energy policy issues among those citizens who have strong environmental attitudes. Similarly, recent research has found that political ideology frames how people view scientific and technological issues, with liberals more trusting of both ecological science and scientists and conservatives much more skeptical (Steel et al. 2006). Certainly this has been a pattern in previous chapters in this book in regard to the issue of climate change.

In a previous study examining the impact of values and policy relevant knowledge on renewable energy support in Oregon and Washington, both values and knowledge variables were the most important predictors of support for alternative energy policies for public survey respondents (Pierce et al. 2009). More specifically, measures of political ideology and the NEP both contributed to perceptions that climate change is a moderate or serious problem requiring policy change to develop renewable energy sources. In addition, those respondents on the left/liberal side were significantly more likely than conservative respondents to support renewable energy development.

That earlier study also found that both subjective and objective energy policy knowledge indicators were positively and significantly correlated with perceptions of climate change and the need for new renewable energy policies. Those respondents with the highest levels of knowledge were significantly more likely to believe that climate change is a moderate to serious problem, and supported policies to encourage the development of renewable sources. Even while controlling for various SES and demographic variables, higher levels of knowledge had an independent effect on perceptions of climate change and renewable policy (Pierce et al. 2009).

A study by Malka, Krosnick and Langer found a strong partisan influence on the relationship of concern for global warming and knowledge levels (2009). Using public surveys conducted in 2006 and 2007, the study found that higher levels of knowledge concerning global warming led to higher levels of concern about the potential impact of global warming among independents and Democrats, but not among Republicans. That study also found that even among Republicans with higher levels of knowledge, there was no relationship of knowledge with their concern about the impact of global warming. Based on the findings of their study, Malka et al., conclude (2009: 645):

> Many people and organizations these days are devoting considerable effort to educating the American public about climate change with the goal of increasing concern about it…Our research suggests that the impact of these efforts will depend upon the predispositions of citizens.

While the Malka et al. study examines the impact of knowledge and partisanship on global warming and not the subject of this chapter—support for AET and positions concerning AET tradeoffs—the previous chapters in this book have found a relationship between concern about climate change and support for AET.

Similar to the Malka et al. study (2009), a Norwegian research project found that "…political party preference has a larger impact on energy technology attitudes than previously believed" (Karlstrom and Ryghaug 2014: 656). The Norwegian study included various socioeconomic and demographic variables, and used the level of education as a proxy measure for knowledge. The results found some impact of the control variables, including those with higher levels of education being more supportive of certain specific AET, but concluded that partisan orientations (e.g., left-right) were more influential.

In an overview of over 30 public opinion polls and studies conducted in the U.K. from 2000 to 2005, McGowan and Sauter (2005) found that knowledge levels varied widely among the public, depending on the type of AET. Public surveys found high levels of knowledge for solar, hydro, and wind energy, but knowledge was relatively nonexistent for biomass energy. McGowan and Sauter also found that the polls and studies they reviewed found higher levels of public knowledge among populations that lived near, and/or had experience with AET such as wind farms (2005, p. 11). They also reported that one study conducted in Scotland found "…a rather strong correlation between knowledge of a specific technology and a positive opinion/general approval' (2005, p. 11).

Researchers at George Mason University and John Hopkins University have conducted public opinion surveys from 2012 to 2015 in the State of Maryland concerning citizen energy preferences and knowledge (Akerlof et al. 2015). While the research team found strong levels of support for the state's Renewable Energy Portfolio Standard (i.e., 20% of electricity generated by renewables by 2022), and support for the expansion of incentives for AET development, public knowledge levels were more mixed. In the 2015 study, only about a third of respondents knew the largest fuel sources used to generate the state's electricity and, even though they supported the state's RPS, only 38% actually knew what the RPS goals were (Akerlof et al. 2015). In addition, while nuclear power accounts for 40% of the state's electricity (tied for the largest power source with coal at 40%), few in any of the state's regions could "correctly identify nuclear as one of the state's largest electricity sources—Western, 5%; Central, 7%; Southern, 15%; Eastern, 11% (Akerlof et al. 2015, p. 14). However, while the Maryland study did not examine the relationship between support for AET development and knowledge levels specifically, it is clear that the Maryland public overwhelmingly supports the development of AET, regardless of knowledge levels.

While the case for a knowledgeable public concerning highly complex scientific and technical issues—such as AET—was discussed in the previous chapter and in the discussion above, one must also heed the warnings of some researchers that notions of public "knowledge deficits" and "public understanding of science" approaches can lead to expert-driven policy-making processes. As Flynn, Bellaby and Ricci argue concerning, the adoption of emergent energy technologies (2011, p. 245):

> Contemporary discourse about public engagement has rejected the 'public understanding of science' (PUS) model, in which lay people were thought to be inadequately informed or irrational about science (or risk) and therefore needed to be 'educated' by experts. The assumption that 'the public' has a 'knowledge deficit' has been challenged, mainly because

it presumes that expert scientific knowledge is always objective and superior to lay knowledge

Similarly, concerning the siting of AET and the typical labeling of opposition as 'NIMBYism,' Devine-Wright argues (2011, p. 60):

> As such, siting fosters an 'information deficit' view of public knowledge in environmental planning and development… presuming local residents to have little or no substantive knowledge regarding ecological or archeological processes. As a consequence, the aim of public engagement in technology siting is likely to be information provision rather than deliberative engagement, only serving to notify residents and communities about the changes that are planned.

While examining levels of public information and knowledge concerning energy in the states studied here, one must remain cognizant of these concerns in how citizens opposed to AETs are often portrayed.

11.3 Analyses

How much do the citizens in this study's surveys think they actually know about alternative energy technologies? While self-assessed familiarity with an issue may be greater than actual knowledge, it also is true that a *sense of knowing* may likewise lead to distinctive behavior or opinion formation. The survey asked respondents the following question:

> How familiar are you with specific renewable energy technologies including biofuel, wind, solar, geothermal and wave energy?

The alternative responses were: 1 = not familiar, 2 = somewhat familiar, 3 = familiar, and 4 = very familiar. We calculated the mean response on familiarity with each AET for each state's respondents. The results are shown in Table 11.1, where the higher score reflects greater self-assessed familiarity. An overall index of familiarity was calculated by summing the familiarity scores across the six

Table 11.1 Average Self-Assessed Familiarity with Alternative Energy Technologies, Controlling for State

	California	Oregon	Washington	Idaho[a]	Total	*p*
Biofuels	2.22	2.23	2.24	2.28	2.24	.60
Wind	2.62	2.62	2.60	2.60	2.61	.93
Geothermal	1.99	1.97	1.96	1.96	1.97	.87
Solar	2.86	2.97	2.77	2.80	2.80	.18
Wave	1.77	1.82	1.77	NA	1.79	.37
Hydro	2.62	2.70	2.80	2.81	2.73	.00
Index	14.1	14.1	14.1		14.1	.95
N =	691	755	710	683	2839	

[a]Idaho respondents were not asked about their familiarity with wave energy since it is a land-locked state

technologies. Because Idaho is a landlocked state without the possibility of marine energy technologies such as wave energy, respondents were not asked their familiarity with wave energy, resulting in a lower summary familiarity index.

Several patterns emerge from Table 11.1. First, overall, residents of the western Pacific states did not consider themselves particularly knowledgeable about AETs. No average in any state reached the level of "familiar." Second, very few differences emerged among the four states. Only on *hydro* was there a statistically significant difference ($p = .001$), where California respondents expressed the least familiarity. This falls in line with the relative use of hydropower as an energy source, given that Washington, Oregon, and Idaho are much more reliant on hydroelectric power than is California. Third, there were large differences among the technologies in terms of the level of familiarity respondents expressed. Thus, solar and hydro had the greatest level of expressed familiarity, while geothermal and wave technologies were least familiar. This pattern surely results from the greater relative public attention paid to solar and hydro across time.

A second approach to the question of the cognitive resources that citizens bring to the policy arena is to examine their ability to correctly answer questions about energy resources in their particular state. Thus, respondents were asked to respond to the following:

(A) The largest source of energy for electricity in (…state…) is:

1. Coal
2. Hydroelectric
3. Natural gas
4. Nuclear

(B) Most electricity in (…state…) is used by the:

1. Residential sector (e.g., households)
2. Commercial sector (e.g., retail stores)
3. Industrial sector (e.g., factories and mills)
4. Transportation sector

(C) Being "off-grid" means:

1. Producing one's own electricity
2. Getting electricity from another state
3. Having no electricity
4. Being energy efficient

Although the correct answers to the questions may vary some among the states (see Chap. 2), it is possible to calculate the scores for each respondent and to sum and average them within each state. The results are shown in Table 11.2, which presents the proportion of "correct" answers in each state for each question, along with the average score for knowledge when summed across the three questions.

In contrast to the sense of familiarity with AETs (Table 11.1), the actual knowledge of a short list of energy questions produced significant differences among the

Table 11.2 Proportion of correct answers to energy questions, controlling for respondents' state

	California	Oregon	Washington	Idaho	Total	*p*
Source	.18	.59	.72	.68	.54	.00
Use	.13	.32	.36	.22	.26	.00
Grid	.65	.62	.61	.60	.62	.22
Index	.96	1.53	1.69	1.51	1.43	.00
N =	693	756	715	687	2851	

populations of the four states. Residents of Washington and Idaho were more likely to correctly identify the predominant source of energy in their state than were those of Oregon, and far more likely than those of California. Washington and Oregon residents were more informed on the primary use of energy than were residents of California and Idaho. On the other hand, knowledge of the meaning of "off the grid" was fairly even across the four states. Overall, the index score shows that the Washington residents were the most knowledgeable, while the Californians were the least well informed.

To be sure, there were significant differences in the results produced by the two indicators of familiarity and knowledge. On self-assessed familiarity with alternative energy technologies, almost no differences emerged among the various state samples. However, the same cannot be said for the measures of actual knowledge about energy. In both measures, there were substantial differences in knowledge level among the particular items. Respondents indicated much greater familiarity with wind, solar, and hydro forms of energy production than with biofuels, geothermal, and wave energy forms, undoubtedly due to the much greater visibility and attention paid to the former. On the actual knowledge questions, overall greater knowledge was evidenced on the "off-the-grid" and source questions than on the greatest use of energy item.

Nonetheless, self-assessed familiarity is not the same as actual knowledge; furthermore, the actual knowledge questions were more general about overall energy production and use, and the familiarity questions were about specific AETs. Thus, a further question emerges, namely what is the relationship of the two measures to each other? Are those respondents who say they know a lot about alternative energy technologies also more likely to actually know more about energy generally? Table 11.3 presents the results of a series of correlations among the different knowledge and energy measures.

While the correlations are not particularly strong, they generally are statistically significant and in a consistent direction. Respondents who believed they were familiar with alternative energy technologies also were more likely to have somewhat greater knowledge of current energy conditions (use, source, grid) in their particular states.

It is probably not surprising that the strongest correlation between source *knowledge* and self-assessed *familiarity* emerged for hydroelectric power generation. Hydro is the major source of energy production in three of the four states, and in the

Table 11.3 Correlations (r) of self-assessed familiarity with alternative energy technologies and actual knowledge about energy conditions

	Knowledge			
Self-Assessed Familiarity:	Source	Use	Grid	Index
Biofuels	.13***	.08***	.24***	.23***
Wind	.06***	.09***	.20***	.18***
Geothermal	.06***	.07***	.25***	.21***
Solar	.01	.07***	.21***	.14***
Wave	.08***	.05**	.21***	.17***
Hydro	.25***	.09***	.23***	.30***
Index	.12***	.08***	.28***	.25***

**p = ≤.01; ***p = ≤.001 (1-tailed); N = 2154–2844

Table 11.4 Regression of knowledge and familiarity indexes on political values, orientations and beliefs

	Knowledge	Familiarity
	b	*b*
Ideology	−.06**	−.07***
NEP	−.11***	−.06*
Postmaterialism	.12***	.14***
Climate Change	−.04	−.07**
Scarcity	−.07***	−.19***
Security	.04	.04*
Technology	−.03	.07**
R^2 =	.03	.08
F =	13.7***	24.3***
N =	2778	2086

*p ≤ .05; **p ≤ .01; ***p ≤ .001

eyes of many residents in the West, because of its long history and its centrality, hydro may also be seen as conventional rather than alternative energy technology.

Overall, then, while the two measures are sufficiently correlated to suggest that there is some element of commonality, there also is enough mutual variation that is not explained by their relationship to suggest continuing to keep them separate. The two measures may be connected to different causal sources, and they may have differential impacts on the respondents' views of alternative energy tradeoffs.

Table 11.4 regresses the two indexes (knowledge and familiarity) on seven political values, orientations, and beliefs, as expressed by the respondents. The political orientations had a greater cumulative impact on familiarity (R^2 = .08) than they did on knowledge itself (R^2 = .03). Liberals expressed somewhat greater familiarity and somewhat more knowledge than did conservatives. On the other hand, respondents high on the New Ecological Paradigm measure were somewhat lower in knowledge and lower in familiarity than were those on the other end of the spectrum. Holding postmaterial values, though, was related to both knowledge and familiarity. The effect of perceived climate change was significant only for the familiarity variable, with those saying yes slightly more likely to indicate familiarity with AETs. Of the

Table 11.5 Regression of knowledge and familiarity on background variables

	Knowledge	Familiarity
	b	*b*
Age	−.05	−.03
Gender	.16***	.24***
Education	.11***	.20***
Length of Residence	−.01	−.03**
Employed	−.03	.01
Income	.05*	−.06*
R^2 =	.06	.10
F =	27.04***	35.90***
N =	2601	1965

*p ≤ .05; **p ≤ .01; ***p ≤ .001

three energy belief dimensions, only scarcity had a significant relationship with both familiarity and knowledge; respondents with greater perceptions of energy scarcity had somewhat lower knowledge and lower expressed familiarity. And, those respondents with a more positive orientation toward technology's role and the importance of energy to security expressed slightly higher levels of familiarity.

As a group, the background variables (Table 11.5) had a greater impact on familiarity than on the knowledge index. Even so, on familiarity, the background variables explained only 10% of the variation. For both knowledge and familiarity, gender and education showed significant positive effects. Men and the more highly educated were more likely to claim familiarity with the AETs and to score higher on the actual knowledge index. When controlling for the other background variables, though, income showed a significant positive relationship with actual knowledge, but a significant negative effect in regard to familiarity.

There are two ways to look at the impacts of knowledge and familiarity on AET tradeoff opinions. The first way is to examine whether those with more knowledge and claimed familiarity have different tradeoff opinions than those with less knowledge and familiarity. That is, does knowing more about energy lead to different tradeoff beliefs when controlling for background attributes and political orientations?

Second, knowledge may not have that direct effect on tradeoff beliefs, but rather it may enhance the ability of an individual to better link those beliefs to political and personal values. The knowledge and/or familiarity with the policy area may facilitate sensitivity to the implications of alternative policy positions for those values. Table 11.6 first looks at the direct impact of knowledge and familiarity, while controlling for selected political orientations and background attributes.

The results shown in Table 11.6 show that when controlling for a range of political and policy attributes, knowledge had a weak direct effect on at most only one of the tradeoff beliefs, namely that concerning hydroelectric power. The direction of that relationship is that greater knowledge is associated with greater support for the position to "continue to operate hydroelectric dams for power even it harms salmon runs."

Table 11.6 Regression of tradeoff beliefs on knowledge, familiarity and selected political orientations and background attributes

	Tradeoff beliefs				
	Index	Wind	Solar	Wave	Hydro
	b	*b*	*b*	*b*	*b*
Knowledge	−.02	−.02	.00	.02	−.06**
Familiarity	.03	.10***	.06*	.03	.00
NEP	.43***	.23***	.30***	.26***	.36***
Ideology	−.06*	.02	−.06**	−.07**	−.03
Scarcity	−.12***	−.09***	−.10***	−.12***	−.02
Climate	−.03	−.02	.03	−.06*	−.04
Gender	−.09***	−.03	−.04	−.11***	−.07**
Education	.00	.03	−.05*	.03	.01
R =	.22	.066	.10	.12	.17
F =	70.1***	16.2***	28.0***	34.0***	51.1***
N =	2686	2689	2699	2693	2.694

*p ≤ .05; **p ≤ .01; ***p ≤ .001

Table 11.7 The correlation of political/policy orientations and AET tradeoff beliefs, controlling for respondents' knowledge level

	Knowledge level			
	0	1	2	3
Ideology	−.05	−.25**	−.29***	−.47***
NEP	.35***	.47***	.41***	.54***
Postmaterialism	.00	.08*	.09**	−.01
Climate Change	−.03	−.22***	−.21***	−.14**
Scarcity	−.07	.07*	.02	.08*
Security	.01	.04	−.07*	−.04
Technology	.00	.01	.03	.01
N=	456	977	957	342

*p ≤ .05; **p ≤ .01; ***p ≤ .001

On the other hand, self-assessed familiarity had a small, but statistically significant impact on two of the tradeoff questions—wind and solar. The effects are in contrasting directions, with greater familiarity leading to greater support for the preservation of tall grass prairie "even if it prevents windmill energy farm production," while on the other hand, it also leads to somewhat greater support for installing "solar power in desert even if it disrupts ecology of endangered turtle."

Even when controlling for the two knowledge variables, the New Ecological Paradigm remained by far the most powerful predictor of the tradeoff attitudes across the board of AETs. To be sure, the perception of scarcity had a significant effect on all but the hydro attitudes, with increased perceptions of scarcity associated with the pro-AET position on the tradeoff questions.

Table 11.7 shows what happens to the impact of political orientations on AET beliefs when we group the respondents into four levels of knowledge holding, based

Table 11.8 Globalist and localist percentage correct on three knowledge questions

Knowledge question	Globalists	Localists
State's largest source of energy	52%	51%
State's largest user of energy	27%	29%
Meaning of off-the-grid	57%	70%
N =	101	470

on the number of correct answers given to the source, use, and off-the-grid questions described earlier in this chapter. The question confronted is whether greater knowledge allows individuals to connect their political orientations and values to their AET beliefs in a way that appears to produce a shared sense of the implications of those beliefs for those values.

Knowledge seemed to have the greatest impact on the relationship of political ideology to the overall tradeoff beliefs index. The greater the knowledge level, the more likely were conservatives and liberals to differ substantially on their AET positions. Thus, well-informed liberals were more likely than well-informed conservatives to take the belief position that the need to install AETs should not trump the need to protect the environment, subject to potential harm by that installation. At the least informed level, the differences between liberals and conservatives are minimal.

A similar pattern holds for the role of knowledge in the impact of the New Ecological Paradigm on AET beliefs, but it was not nearly so strongly as for political ideology. Moreover, even for the uninformed, the NEP and DSP advocates differed substantially on AET beliefs. While there were other scattered significant correlations across the matrix in Table 11.7, none were as consistent as for ideology and the NEP. Thus, for perceived climate change, the lowest level of knowledge showed the lowest impact on AET belief, but the highest knowledge level had correlations lower than the two middle positions. And, likewise for postmaterial values, significant correlations resulted at the two middle levels.

It is clear that the NEP had the strongest correlation with AET beliefs at each of the knowledge levels. But, how did the globalists (high NEP/pro-AET) and localists (high NEP/anti-AET) differ in knowledge levels? Table 11.8 shows the percentage of correct answers to each of the three knowledge questions for the globalists and localists. On two of the questions (the greatest state source of energy and the greatest state use of energy) there were no differences between the two groups. On the off-the-grid questions, though, the localists were more often correct (70%) than were the globalists (57%).

What people know about the energy policy area did make a difference in how they connected their fundamental political and environmental values to their positions on the AET tradeoff questions. Perhaps because the path from alternative energy to political ideology is less direct than it is for the path to NEP values, the impact of knowledge is even greater in the former. So, it is not that knowledge brings potential disputants together in their relationship to an NEP-relevant policy question, but rather it brings people who share a political or environmental value closer together on those questions, but farther apart from those with whom they disagree politically.

11.4 Conclusion

The findings in this chapter support the thesis that public awareness and knowledge can lead to enhanced public support for the efforts needed to mitigate climate change through the use of AET. Because everyday citizens are either directly or indirectly involved in the policy process and engage in activities that contribute to climate change, it is indeed important to assess the scope and depth of their policy-relevant knowledge and to flesh out the strength of the link between policy-relevant knowledge, support for the development of AET, and AET tradeoffs. At the same time, however, opposition to AETs by local environmentalists on behalf of immediately-affected elements of the environment suggest that by more clearly specifying the connection between knowledge holding, attitudes toward climate change and energy, and the importance of local environmental effects, purposeful public education and information dissemination efforts can be targeted more effectively.

References

Akerlof, K., Winch, P., Parker, C., & Buckland, A. (2015). *Public knowledge, behaviors, and preferences about energy: A Maryland statewide survey, fall 2015*. Fairfax: Center for Climate Change Communication, George Mason University.

Ansolabehere, S., & Konisky, D. M. (2014). *Cheap and clean: How Americans think about energy in the age of global warming*. Cambridge, MA: MIT Press.

Bolson, T., & Cook, F. L. (2008). The polls: Public opinion on energy policy: 1974–2006. *Public Opinion Quarterly, 72*, 364–388.

Cobb, K. (2016, March 27). Energy policy and uninformed opinion. *Resource Insights.*. http://resourceinsights.blogspot.com/2016/03/energy-policy-and-uninformed-opinion.html. Accessed 15 Aug 2016.

Dahl, R. A. (1985). *Controlling nuclear weapons: Democracy versus guardianship*. Syracuse: Syracuse University Press.

Delli Carpini, M., & Keete, S. (1996). *What Americans know about politics and why it matters*. New Haven: Yale University Press.

Devine-Wright, P. (2011). From backyards to places: Public engagement and the emplacement of renewable energy technologies. In D. Devine-Wright (Ed.), *Renewable energy and the public: From NIMBY to participation*. Earthscan: New York.

Egan, M. (2016, May 27). Donald Trump's energy plan: Regulate less, drill more. *CNN Money*. http://money.cnn.com/2016/05/26/investing/donald-trump-energy-plan/. Accessed 15 Aug 2016.

Ettema, J. S., & Kline, F. G. (1977). Deficits, differences and ceilings: Contingent conditions for understanding the knowledge gap. *Communication Research, 4*, 179–202.

Flynn, R., Bellaby, P., & Ricci, M. (2011). The limits of upstream engagement in an emergent technology: Lay perceptions of hydrogen energy technologies. In D. Devine-Wright (Ed.), *Renewable energy and the public: From NIMBY to participation*. Earthscan: New York.

Genova, B. K., & Greenberg, B. S. (1979). Interest in the news and the knowledge gap. *Public Opinion Quarterly, 43*, 79–91.

Jamieson, K. H. (2000). *Everything you think you know about politics…and why you're wrong*. New York: Basic Books.

Janicke, M. (1997). The political system's capacity for environmental policy. In M. Janicke & H. Weidner (Eds.), *National environmental policies: A comparative study of capacity-building*. New York: Springer.

Karlstrom, H., & Ryghaug, M. (2014). Public attitudes towards renewable energy technologies in Norway. *Energy Policy, 67*, 656–663.

Malka, A., Krosnick, J. A., & Langer, G. (2009). The association of knowledge with concern about global warming: Trusted information sources shape public thinking. *Risk Analysis, 29*, 633–647.

McGowan, F., & Sauter, R. (2005). *Public opinion on energy research: A desk study for the research councils*. Brighton: The Freeman Centre, University of Sussex.

Pew Research Center. (2008, March 6). Public sends mixed signals on energy policy. *Pew Research Center Publications*. http://pewresearch.org/. Accessed 15 Aug 2016.

Pierce, J. C., & Lovrich, N. P. (1986). *Water resources, democracy and the technical information Quandary*. Millwood: Associated Faculty Press.

Pierce, J. C., Lovrich, N. P., Tsurutani, T., & Abe, T. (1989). *Public knowledge and environmental politics in Japan and the United States*. Boulder: Westview Press.

Pierce, J. C., Steger, M. A., Steel, B. S., & Lovrich, N. P. (1992). *Citizens, political communication, and interest groups*. Westport: Praeger Publishers.

Pierce, J. C., Lovrich, N. P., Steel, B. S., Steger, M. A., & Tennert, J. (2000). *Political culture and public policy in Canada and the United States: Only a border apart?* New York: Edwin Mellen Publishers.

Pierce, J. C., Steel, B. S., & Warner, R. L. (2009). Knowledge, culture and public support for renewable energy technology policy in Oregon. *Comparative Technology Transfer and Society, 7*, 270–286.

Smith, E. (2002). *Energy, the environment, and public opinion*. Oxford: Rowman and Littlefield.

Steel, B. S., Soden, D., & Warner, R. L. (1990). The impact of knowledge and values on perceptions of environmental risk to the Great Lakes. *Society and Natural Resources, 3*, 331–348.

Steel, B. S., Lach, L., & Satyal, V. (2006). Ideology and scientific credibility: Environmental policy in the American Pacific Northwest. *Public Understanding of Science, 15*, 481–495.

Chapter 12
Information Source Reliance and Alternative Energy Tradeoffs

10 percent of Americans take a "dismissive" attitude toward climate science and think climate change isn't happening. Why the gap? Misleading representations of climate science in the media have a lot to do with it. (Aaron Huertas, Press Secretary, Union of Concerned Scientists; Huertas 2012)

Some mainstream media around the world have a tendency to publish misinformed or, worse, systematically and falsely negative stories about renewable energy. Some of those stories' misinformation looks innocent, due to careless reporting, sloppy fact checking, and perpetuation of old myths. But other coverage walks, or crosses, the dangerous line of a disinformation campaign—a persistent pattern of coverage meant to undermine renewables' strong market reality. (Armory B. Lovins, Chief Scientist, Rocky Mountain Institute; Lovins 2013)

12.1 Introduction

It is known that democratic citizens vary widely in the level of knowledge they hold, as well as in what they profess to be true (Delli Carpini and Keeter 1996). This book has argued throughout that the level of knowledge is especially significant in scientific and technical policy areas. Those policy questions ordinarily are not part of the daily discourse of the "average American." This work suggests that as the pressures for increased citizen influence and involvement in those policy arenas increase, the "technical information quandary" takes center stage. That is, how are citizens to make informed policy judgments reflecting their values and self-interest when they may not hold the knowledge required to inform those judgments? It also is known that there is wide variation in the information content to which individuals are exposed, both across time and across information sources.

The reality of climate change is one arena in recent years where broadly held knowledge and policy view gulfs are evident among significant segments of political and social structures and systems. Thus, questions confronted in this chapter

J.C. Pierce, B.S. Steel, *Prospects for Alternative Energy Development in the U.S. West*, Environmental Challenges and Solutions 4,
DOI 10.1007/978-3-319-53414-5_12

include: what information sources do the citizens of California, Oregon, Washington and Idaho say they use "to learn more about California's (e.g.) energy situation and policy." And, what are the implications of reliance on those information sources for policy-relevant knowledge and alternative energy technology (AET) preferences? Are there variables that may account for any differences in reported information sources, such as state of residence, demographic attributes, and political/policy orientations? The analysis looks at the relationship of reported information-source dependence with the public's level of knowledge in regard to energy, as well as reported familiarity with AET. Lastly, the chapter examines the relationship of reported information source dependence with beliefs about the AET tradeoffs, controlling for background, political orientations, and knowledge.

12.2 Information Source Reliance

The literature on information source reliance suggests that citizens use a variety of sources for policy-relevant knowledge. According to the Gallup Poll, the American public's most frequently used source of information is typically television (e.g., local and national news programs), even though cynicism concerning the reliability of information provided by this medium is widespread (Pierce et al. 1992; Steel et al. 1992). In 2013, Gallup found that 55% of American used television for their main source of news concerning current events, followed by 21% for the Internet, 9% for newspapers and 6% for radio (Saad 2013). When examining information sources used by the public while controlling for various demographic and socioeconomic status (SES) characteristics, Gallup found that all age groups reported television as the main source of information, but that older cohorts were most likely to use television, and were much more likely to read a newspaper, compared to younger cohorts. In addition, college graduates and post-graduates were significantly less likely than high school graduates to use television and more likely to use the Internet (Saad 2013). Post-graduates were also found to be more likely to use print media for information (19%), compared to other educational groups (7–8%).

The newest medium of information diffusion is the Internet, where the range of possible information sites is nearly countless. Use of the Internet as an information source has increased dramatically in recent years, often at the expense of newspaper readership (Lupia and Baird 2003). While some Internet sites are highly informative and useful sources of policy-relevant information, others are unreliable at best. It is commonly observed by researchers that "the amount of bad stuff out there is truly staggering" (2000, p. 4). It would appear that the Internet is rather like all other information sources—they can either inform or misinform users (Bimber 1998; Lupia and Baird 2003; Uslaner 2004).

In terms of social media, a recent study by the Pew Research Center found that 30% of Americans use Facebook for news, 10% use YouTube, and 8% use Twitter (Anderson 2014). Of those who use social media for news and information, 50% either shared or reposted news stories or videos, 46% discussed a news issue or event, 14% posted photos they took of news events, and 12% said they post videos of a news event during the year (Anderson 2014). However, when Pew asked Facebook users what type of "news" they regularly view, 73% said "entertainment" news, 65% said "people and events in my community," and 57% said "sports" news. The fourth most frequently used type of news encountered on Facebook was "national government and politics" (55%). Other types of news mentioned included "local government and politics" (44%), "international news" (39%), and "science and technology" (37%).

While some observers have been concerned about the use of social media for information by younger cohorts, a recent study by the American Press Institute concerning "millennials" (18–34 years) and their use of the Internet and social media interest in news found that (American Press Institute 2015, p. 1) "this newest generation of American Adults is anything but "newsless," passive, or civically uninterested… Millennials consume news and information in striking different ways than previous generations, and their paths to discovery are more nuanced and varied than some may have imagined."

The study further found that millennials regularly read a mix of traditional news, practical "news you can use," and local community news, and follow various social issues—often five or more news topics at a time (American Press Institute 2015, p. 2). Of course, the credibility of the information encountered on social media as well as other more traditional sources can be questionable and require additional research (e.g., Hesse et al. 2005; Westerman et al. 2014).

As to the impact of particular information source reliance on policy relevant knowledge, some researchers have found that "exposure to national television news was not related to any of the knowledge domains" (Jamieson 2000, p. 152). Similarly, research by Steger et al. (1988) and Pierce et al. (1992) found a negative relationship between television use and environmental knowledge in the United States. In contrast, this same research effort revealed that the frequency of newspaper reading was strongly correlated with policy-relevant knowledge holding concerning the environment. This finding is consistent with other research showing a positive relationship between newspaper readership and knowledge holding in general (Jamieson 2000). It is likely that a reciprocal relationship is present—that is, more knowledgeable individuals are more inclined to read newspapers frequently, and the reading of newspapers leads to the further acquisition of policy-relevant knowledge.

As with other information sources, the radio can be used for a diverse set of purposes ranging from the transmission of educational programming such as that often featured on National Public Radio (NPR) to the airing of various "talk shows" catering to ideologues. Previous research reported by Steger et al. (1988) found that in the Canadian context radio listening was positively correlated with knowledge of acid rain, but in the American context it was negatively correlated with the same knowledge indicators. Other research suggests only a slightly positive correlation of radio use with the holding of general political knowledge (Delli Carpini and Keeter 1996).

According to Malka, Krosnick and Langer, "many people and organizations these days are devoting considerable effort to educating the American public about climate change with the goal of increasing concern about it" (2009, p. 645). However, they also found in their study of the American public that "trust in information sources" is mediated by partisanship, with Democrats and Independents more trusting of information from scientists concerning global warming and Republicans less trusting. They conclude their study by commenting (2009, p. 645). "Our research contributes to the growing literature on trust in information sources generally and on trust in scientists in particular. Decades of research in social psychology show that people are more receptive to the views of information sources they trust...."

In prior research examining information-source reliance and support of renewable energy technologies, four different information sources (television, radio, the Internet, and newspapers) were examined as predictors of knowledge of energy policy issues (Pierce et al. 2009). The findings of that study revealed that the Internet and newspapers had statistically significant impacts on energy policy knowledge. Frequent use of the Internet was found to be related to higher levels of energy policy knowledge and frequent use of newspapers was also positively and significantly related to higher energy policy knowledge scores. However, neither television nor radio use had a significant effect in energy policy knowledge levels.

The use of multiple sources of information has also been found to produce more favorable outcomes in terms of policy understanding and participation in policy processes. Research by Sotirovic and McLeod found the following (2001, p. 273):

> ...individuals' efforts to think about news and search for additional information and perspectives modify what people "get from" media. Reflecting about news and integrating from various sources promote better understanding of the political world and may provide a stronger cognitive base for political participation than factual knowledge.

The following section examines where citizens in California, Oregon, Washington, and Idaho obtain energy-related information, and how information from media sources, along with trans-situational and cultural factors, affect levels of understanding. The analysis looks at knowledge both in terms of what people *say* they understand (subjective knowledge) as well as whether or not they can *correctly answer* questions about energy (objective knowledge).

12.3 Analyses

Where do the respondents in this study's four-state survey say they obtain their information about energy situations and policy in their own state? The answers to that question are important to the degree that one can suggest that alternative sources may provide different information, different takes on information, or different contexts within which to interpret that information. Respondents to the survey were asked to indicate the degree to which they used a number of alternative sources, the

Table 12.1 Public information source reliance by source and by state

Source	Total	California	Oregon	Washington	Idaho	F	Sig
Television	2.85	2.79	2.89	2.85	2.89	2.35	.07
PBS/CBC	2.53	2.51	2.58	2.52	2.50	.96	.41
Radio	2.31	2.26	2.29	2.34	2.35	1.42	.24
Major paper	2.58	2.48	2.66	2.54	2.63	5.11	.00
Other paper	2.30	2.24	2.32	2.30	2.34	1.49	.22
Local leaders	1.81	1.74	1.79	1.84	1.87	4.49	.00
State elected	1.86	1.78	1.85	1.89	1.92	4.34	.00
State energy	1.95	1.97	1.89	2.00	1.94	2.50	.06
Univ/College	1.98	1.99	1.94	2.01	1.98	.83	.48
Utilities	2.37	2.35	2.33	2.42	2.36	1.68	.17
Env. groups	2.03	2.04	2.07	2.01	1.98	1.34	.26
Internet	2.61	2.66	2.55	2.63	2.62	1.52	.21
Total[a]	27.15	26.8	27.1	27.3	27.4	2.1	.10
Source diversity[b]	.73	.75	.76	.70	.72	.13	.01

N = 2829–2840

[a]Total is average summation of source reliance across individuals. Scheffe tests showed no significant differences among the states

[b]Source diversity is the average individual level variance among reliance on the sources. Scheffe tests show a significant difference between Oregon and Washington publics

response options of never, infrequently, frequently, and very frequently. The potential information sources are listed in Table 12.1. That table also shows the average response of the public in each of the states for each of the potential information sources. A higher score indicates more frequent use of the source, with 4 the maximum score. There are two ways to look at the results—the first is the differences among the states' publics, and the second is the differences among the information sources.

Overall, California respondents were somewhat less likely than the publics of the other three states to indicate reliance across the full range of information sources in regard to their state's energy situation and policy (p = .10). Even so, those differences were not large, and are only marginally significant in statistical terms. Californians were less likely to use major newspapers (p = .002), local leaders (p = .004), and state elected officials (p = .005). Perhaps more interesting are the differences among the information sources themselves. Clearly, the dominant source of information across all four states was television, followed by the Internet, "major newspapers," and PBS. At the other end of the spectrum, that of less frequent use, were local leaders, state elected leaders, and state energy officials. Of the frequent combatants on energy policy, in-state environmental groups were less often used as a source than were energy utilities. The publics of California and Oregon showed greater variance among information sources in their reliance on those sources, suggesting some possible selective-source attention.

A second question is whether there are groupings or patterns of information source reliance across the different sources, such that individuals who relied on a

Table 12.2 Factor loadings (varimax rotation) of information source reliance

	Factor			
Source	I	II	III	IV
TV news	.23	.32	.15	**−.63**
PBS	.14	**.82**	.07	−.01
Radio	.00	**.75**	.06	.07
Major newspaper	.07	.17	**.80**	−.03
Local newspaper	.11	−.02	**.85**	.08
Local leaders	**.70**	.07	.26	−.02
Elected officials	**.80**	.05	.14	−.06
State energy	**.77**	.08	.00	.15
Universities	**.51**	.15	.07	**.53**
Utilities	**.63**	.05	−.04	.07
Environmental groups	.37	.36	.09	**.47**
Internet	.09	.10	.07	**.72**
% Variance	27%	12%	10%	9%

N = 2829–2840

particular source were also more likely to rely on certain other sources. If members of the public are selective in their source reliance based on their policy or political orientations, then we may find a clue as to how they see the sources in particular groups. In order to answer this question, the analysis employs a factor analysis of the responses to the multiple information sources. The results are shown in Table 12.2.

The results of the factor analysis show that there were distinct (and understandable) patterns in the reliance of the public on information sources about energy policy. Thus, individuals who relied on local leaders also tended to rely on elected officials, the state energy office, universities, and utilities, but less so on the other sources. These four information sources share a kind of political authority and/or expertise in regard to providing information about energy policy. This factor by far introduces the greatest explained shared variation, suggesting it is the dominant and more central (but clearly not the only) factor. Public Broadcasting and radio, expressing the common content to the two aural media, share the second factor. Likewise, as shown in Factor III, respondents who relied on major newspapers also tended to rely on local newspapers, but showed no distinct pattern in regard to the other sources. Factor IV loadings produced an intriguing pattern. Respondents who relied on environmental groups also tended to rely on universities and the Internet. On the other hand, television news has a *negative* loading on this factor, suggesting a substantial gulf in the type of information source on which they depend. That is, those who look to environmental groups, universities and the Internet for information, tended not to look to television news.

The next question was whether different types of respondents, in terms of their demographic attributes, were more or less likely to rely on particular types of information sources for information about energy policy and situations. Thus, six

Table 12.3 Correlations (r) of demographic attributes with type of information source reliance

	Demographic attribute					
Information sources	Age	Gender	Education	Residency	Employed	Income
Official	.12***	−.04*	.01	.12***	.08***	−.01
PBS/Radio	.03	−.01	.11***	.01	.00	.10***
Newspapers	.06**	.00	.04*	.02	.02	.00
Internet/Universities	−.36***	−.04*	.28***	−.28***	−.26***	.13***

*p ≤ .05; **p ≤ .01; ***p ≤ .001; N = 2707–2834

personal attributes (age, gender, education, residency, employment, and income) were correlated with each respondent's score on each of the four factors. Those factor scores represent the degree to which the respondent fit the overall pattern for that particular factor. The results are shown in Table 12.3.

Each *type* (factor) of energy information source expresses a significant correlation with two or more of the demographic attributes shown in Table 12.3. Thus, "official" sources were more likely to be used by respondents who were older, female, had lived longer in the state, and were employed. The respondents who relied more on PBS and radio reported higher levels of education and income. Those who relied on newspapers were somewhat older and somewhat better educated. The strongest relationships emerged for the dimension having to do with universities, the Internet, and environmental groups. These sources were more likely to be used by respondents who were younger, female, more highly educated, newer residents, less likely to be employed, and with slightly higher incomes. This last set of relationships surely is consistent with research on the adoption of new information technologies (e.g., Chan-Olmsted et al. 2013; Nguyen and Western 2007), such as the Internet, among younger members of the public. Likewise, one would expect more highly educated individuals to be more comfortable with and trusting of university sources of information.

It also is important to understand the relationship between information source reliance and individuals' political and policy values and beliefs. Earlier chapters have shown that those values and beliefs are sometimes connected to the positions taken by respondents on the AET tradeoff questions. To be sure, once again these relationships may be reciprocal. One may choose a certain type of information source because of the perception that it will be more likely to provide information relevant to one's values, or supportive of one's values, so that those who share values would share that reliance. On the other hand, patterns of information reliance, perhaps as a consequence of the demographic attributes just addressed, may produce shared beliefs among those who rely on a particular type or set of sources. Clearly this relationship speaks to many contemporary observations about the different policy positions alleged to be represented in different media and different types of media. This analysis looks at the effect on information source reliance of liberal/conservative political orientation, the New Ecological Paradigm, postmaterialism, the perception of climate change, and the perception of scarcity, security, and technology as central to contemporary energy issues. The results are shown in Table 12.4.

Table 12.4 The relationship (r) of political values and information source reliance

	Sources			
Values	Officials	PBS	Newspapers	Internet/University
Liberal/Conservative	.05**	−.19***	−.03	−.21***
NEP	.00	.17***	.02	.12***
Postmaterialism	−.06***	.12***	−.01	.16***
Climate change	−.04*	−.18***	−.05**	−.07***
Scarcity	.08***	.08***	−.05**	−.01
Security	.03	−.01	.01	−.07**
Technology	.16***	.05**	−.04*	.02

*p ≤ 05; **p ≤ .01; ***p ≤ .001; N = 2825–2838

Table 12.5 Correlation (r) of information source reliance factor scores with energy knowledge and renewable energy familiarity

Information reliance factor Score	Knowledge	Familiarity
Officials	.00	.08**
PBS/Radio	.05**	.14***
Newspapers	−.01	.13***
Internet/environmental groups/ universities	.14***	.30***

*p ≤ .05; **p ≤ .01; ***p ≤ .001; N = 2824–2833

Liberals were more likely than conservatives to say that they relied on PBS, universities, and the Internet, reflecting descriptions of those connections in much of the popular press, especially from more conservative sources. Similarly, individuals with greater support for the NEP were also more likely to rely on PBS, universities, and the Internet. The same was true for postmaterialist value holders, although they also claimed to be less likely to rely on officials for information. Those who agreed that climate change is occurring were more likely to rely on information sources across the board, but the strongest relationship was for reliance on PBS. Those who saw scarcity of energy resources as a problem relied more on officials and PBS and somewhat less on newspapers. Those who saw security as an important energy issue were less likely to rely on the Internet and universities. Those who saw technology as a solution to energy policy challenges were more likely to rely on officials and PBS and less likely to rely on newspapers. In short, then, political values and policy beliefs appear to have a strong role in the sources on which the public relies for information about energy situations and policy questions. Of those values and beliefs, the strongest impact appears to come from the liberal-conservative political orientation and the NEP.

The next question to be addressed is whether information-source reliance is related to an individual's knowledge about his/her state's energy use and his/her self-proclaimed familiarity with specific renewable energy technologies. As Table 12.5 shows, actual knowledge was associated with the use of PBS/radio sources of information, but was much more strongly associated with the use of the

Table 12.6 The relationship (r) of Information Source Reliance Scores and Alternative Energy Tradeoff Beliefs

	Information source			
AET Tradeoff	Officials	PBS	Newspapers	Internet/universities/environmental groups
Wind	.00	.05**	.07**	.03
Solar	−.01	.08***	.01	.06**
Wave	.08***	.07***	.03	.11***
Hydro	−.02	.13***	.02	.06**
Summary index	.02	.12***	.04*	.09***

*$p \leq .05$; **$p \leq .01$; ***$p \leq .001$; N = 2761–2771

Internet, environmental groups, or university sources. Self-assessed familiarity with renewable energy technologies, however, had significant associations with all four of the information-source reliance scores, and in each case the correlations were substantially higher than was the case for actual knowledge.

Once again the question can be raised as to the causal direction of those relationships. Thus, does greater reliance on the Internet, environmental groups, and universities lead to a greater sense of familiarity with renewable energy technologies? Or, on the other hand, are those who feel they are familiar with technologies more likely to turn to those particular information sources, perhaps because of some sense of greater congruity in the values in the content?

The same kind of reciprocal relationship might be present in terms of the connection of information source reliance and AET tradeoff beliefs. Do individuals who rely more heavily than others on particular kinds of information sources have different kinds of tradeoff beliefs? To provide a preliminary answer to that question, Table 12.6 shows the correlations between those beliefs and the information source reliance factor scores.

Reliance on public radio and on the Internet exhibited the most consistent statistically significant relationships with beliefs in regard to alternative energy tradeoffs. To be sure, the effect magnitudes are not particularly powerful. On the other hand, the patterns are consistent, namely those who relied on public radio and the Internet were more likely to oppose the siting of AETs if they would have relatively short-term negative environmental effects. Moreover, do these relationships hold up when simultaneously taking into account background attributes and political and social values? In a preliminary look at that question, Table 12.7 presents the results of a multiple regression analysis that examines the individual effects on AET tradeoff beliefs of political orientation, the New Ecological Paradigm, self-assessed familiarity, age, education, reliance on PBS, and reliance on the Internet. When controlling for the other variables, two seem to be clearly the dominant: the NEP values clearly continue to be a primary determinant of AET beliefs. And, the most significant of all is reliance on environmental groups for information, even while controlling for environmental values (i.e., NEP).

As in the prior survey-based chapters, one additional question remains to be answered, namely how do the globalists and localists differ in terms of reliance on

Table 12.7 Multiple regression of summary AET beliefs on selected values, familiarity, background attributes and information sources

Variables:	b
Political orientation	−.11
NEP	.21***
Environmental group reliance	.44***
Age	.00
Education	.03
PBS reliance	.05
Internet reliance	−.13*
R^2 =	.22
F =	104.6***
N =	2667

* ≤ .05; ** ≤ .01; *** ≤ .001

Table 12.8 Globalist and localist differences in reliance on information sources for energy policy and conditions information (T-tests)

	Globalists	Localists	Sig.
Energy information source	Mean	Mean	
Television news	2.73	2.73	.99
Public broadcasting	2.53	2.74	.03
Radio	2.29	2.44	.17
Major newspapers	2.43	2.72	.01
Local newspapers	2.09	2.39	.01
Community leaders	1.59	1.81	.01
State elected officials	1.63	1.83	.01
State department of energy	1.88	1.96	.40
Universities and colleges	1.93	2.06	.16
Utilities	2.33	2.30	.78
Environmental groups	2.03	2.53	.00
Internet	2.59	2.66	.58

N = 569–571

various information source about energy policy? Recall that these two groups shared strong environmental values, but differed in terms of their support for AETs when those technologies had local negative environmental effects. The answer to the question of information-source reliance among globalists compared to localists is shown in Table 12.8. There were some significant differences in the sources from which the two groups obtained information about energy policy and conditions. Generally speaking, localists were more likely ($p = .05$) to rely on many of the information sources. Within that context, compared to globalists, localists claimed to be more reliant on (1) public radio, (2) major newspapers, (3) local newspapers, (4) community leaders, (5) state elected officials, and (6) environmental groups. The unanswered question may be causal: do the localists have their particular perspective because of the sources on which they rely, or do they rely on those sources because of their more localist perspective? The answer is likely a combination of both their values and their information sources, with a spiraling interaction of looking to environmentalists for their information because, on the one hand, of the correspondence of their values and those of the environmentalists, and, on the other hand, because information from environmentalists also reflects the values of that source.

12.4 Conclusion

This chapter examined the sources from which California, Idaho, Oregon and Washington residents report that they obtain information on energy policy issues, as well as what impact other variables—e.g., demographic attributes and political/policy orientations—have on that information-source reliance. This analysis also examined the relationships of reported information source dependence to beliefs about the alternative energy tradeoffs, controlling for background, political orientations, and knowledge levels.

When examining information source use, respondents across all four states indicated that the dominant source of information was television news, which was followed closely by the Internet, major newspapers, and PBS. However, Californians were less likely to use major newspapers, local community leaders, and state elected officials as information sources than were respondents in Oregon, Washington, and Idaho. Also, environmental groups were less often used as information sources than were energy utilities for energy policy. Moreover, the publics of California and Oregon showed a greater willingness to use more variety of information sources compared to the publics of Idaho and Washington.

Respondents who relied on local community leaders for information also tended to rely on elected officials, the state energy office, universities, and utilities. These four information sources are based on notions of political authority and/or expertise in regard to providing information about energy policy. Respondents who relied on environmental groups for information also relied on universities and the Internet as information sources, but they *did not* rely on television news, suggesting a possible preference for nontraditional and/or elite-challenging information sources.

Similar to research findings reported in other studies, political values and ideology also play a role in the types of information sources the public uses. Liberals were found to be more likely than conservatives to rely on public television, universities, and the Internet for energy information. Similarly, individuals that had higher scores for the NEP were also more likely to rely on public television, universities, and the Internet. The same was also found for those respondents with postmaterialist values, although they also claimed to be less likely to rely on government officials for information, compared to those with materialist value orientations.

In terms of AET tradeoffs, those who relied on public radio and the Internet were more likely to oppose the siting of those technologies because of perceived negative local environmental effects. Once again, this is consistent with other patterns evident in this chapter, in that those identifying with the NEP were also most likely to use PBS and the Internet. The implications of these findings are consistent with the observations of Malka et al. (2009), namely that the source of information for energy policy knowledge is an important consideration when examining knowledge levels and policy preferences, but that knowledge acquisition also is mediated by other values, such as political orientation and environmental values. By more clearly specifying the connection between information sources, knowledge holding, and attitudes toward AET, purposeful public education and information dissemination efforts can be targeted more effectively.

Finally, localists were more likely than globalists to use most of the information sources; the largest differences between the two polar groups were in reliance on environmental groups for information. The localists were more concerned about the immediate negative environmental impacts of the installation of AETs, and they were likely to obtain information about those negative environmental effects from environmental groups; however, one can suggest that perhaps they access information from those environmental groups because they already are very pro-environment in their values. At the same time, though, there surely are other reasons for the differences in the choices made by globalists and localists any of which are addressed in the preceding chapters.

References

American Press Institute. (2015). *How millennials get news: Inside the habits of America's first digital generation.* https://www.americanpressinstitute.org/publications/reports/survey-research/millennials-news/. Accessed 16 Aug 2016.

Anderson, M. (2014, September 14). *How social media is reshaping news.* Pew Research Center. http://www.pewresearch.org/fact-tank/2014/09/24/how-social-media-is-reshaping-news/. Accessed 13 Aug 2016.

Bimber, B. (1998). The internet and political transformation: Populism, community, and accelerated pluralism. *Polity, 31*, 133–160.

Chan-Olmsted, S., Rim, H., & Zerba, A. (2013). Mobile news adoption among young adults: Examining the roles of perceptions, news consumption, and media usage. *Journalism and Mass Communication Quarterly, 90*, 126–147.

Delli Carpini, M., & Keeter, S. (1996). *What Americans know about politics and why it matters.* New Haven: Yale University Press.

Hesse, B. W., Nelson, D. E., Kreps, G. L., Croyle, R. T., Arora, N. K., Rimer, B. K., & Viswanath, K. (2005). Trust and sources of health information: The impact of the nternet and its implications for health care providers. *JAMA Internal Medicine, 165*, 2618–2624.

Huertas, A. (2012, October 9). *Why you need to worry about how the media covers climate change.* Renewable Energy World. http://www.renewableenergyworld.com/articles/2012/10/why-you-need-to-worry-about-how-the-media-covers-climate-change.html. Accessed 13 Aug 2016.

Jamieson, K. H. (2000). *Everything you think you know about politics ... And why you're wrong.* New York: Basic Books.

Lovins, A. B. (2013, July 31). *Debunking the renewables "disinformation campaign."* Rocky Mountain Institute. http://blog.rmi.org/blog_2013_07_31_debunking_renewables_disinformation_campaign. Accessed 13 Aug 2016.

Lupia, A., & Baird, Z. (2003). Can web sites change citizens: Implications of web white and blue 2000. *Political Science and Politics, 37*, 77–82.

Malka, A., Krosnick, J. A., & Langer, G. (2009). The association of knowledge with concern about global warming: Trusted information sources shape public thinking. *Risk Analysis, 29*, 633–647.

Nguyen, A., & Western, M. (2007). Socio-structural correlates of online news and information adoption/use: Implications for the digital divide. *Journal of Sociology, 43*, 167–185.

Pierce, J. C., Steel, B. S., & Warner, R. L. (2009). Knowledge, culture and public support for renewable energy technology policy in Oregon. *Comparative Technology Transfer and Society, 7*, 270–286.

Pierce, J. C., Steger, M. A., Steel, B. S., & Lovrich, N. P. (1992). *Citizens, political communication, and interest groups*. Westport: Praeger Publishers.

Saad, L. (2013, July 8). TV is Americans' main source of news. *Gallup Poll*. http://www.gallup.com/poll/163412/americans-main-source-news.aspx. Accessed 13 Aug 2016.

Sotirovic, M., & McLeod, J. M. (2001). Values, communication behavior, and political participation. *Political Communication, 18*, 273–300.

Steel, B. S., Lovrich, N. P., & Pierce, J. C. (1992). Trust in natural resource information sources and post-materialist values: A comparative study of U.S. and Canadian citizens in the Great Lakes area. *Journal of Environmental Systems, 22*, 123–136.

Steger, M. A., Pierce, J. C., Steel, B. S., & Lovrich, N. P. (1988). Information source reliance and knowledge acquisition: Canadian/U.S. comparisons regarding acid rain. *Western Political Quarterly, 41*, 747–764.

Uslaner, E. (2004). Trust, civic engagement, and the internet. *Political Communication, 21*, 223–242.

Westerman, D., Spence, P. R., & van Der Heide, B. (2014). Social media as information source: Recency of updates and credibility of information. *Journal of Computer-Mediated Communication, 19*, 171–183.

Chapter 13
What Has Been Learned? Process and Values Matter!

Abstract This concluding chapter summarizes what has been learned from this work's case studies and the surveys of citizens in the western U.S. As discussed in the introductory chapter, the three Pacific Coast states of Washington, Oregon and California have been active participants in the Pacific Coast Action Plan on Climate and Energy. These three states are host to numerous alternative energy technology (AET) installation controversies; each state has a long history of environmental concerns and has spawned major home-grown environmental movements; and each state has extensive experience with both carbon-based and non-carbon-based (such as hydroelectric) energy production for the power grid. The state of Idaho is included in the analyses as the case of a neighboring state that has not been so proactive promoting non-hydro AET. The case studies demonstrated that the process of siting AET should be transparent, fair, and incorporate local concerns about the local ***environmental*** consequences of AET siting. The case studies and survey data also reveal that ***values***—including political orientations, postmaterialism, and most importantly the NEP—are important predictors of support for renewable energy technologies, but also play a role in predicting public preferences concerning environmental tradeoffs. At the same time, this study also found that globalists (strong environmental values with strong support for AET over local environmental concerns) differ in certain of their values from localists (strong environmental values with strong support for local environmental concerns over AET).

Keywords Alternative Energy Technology (AET) • Energy policy • New ecological paradigm • Political orientations • Values

13.1 Introduction

The Republican Party is committed to domestic energy independence. The United States and its neighbors to the North and South have been blessed with abundant energy resources, tapped and untapped, traditional and alternative, which are among the largest and most valuable on earth. Advancing technology has given us a more accurate understanding of the nation's enormous reserves that are ours for the development. The role of public officials must be to encourage responsible development across the board. Unlike the current Administration, we will not pick winners and losers in the energy marketplace. Instead, we

J.C. Pierce, B.S. Steel, *Prospects for Alternative Energy Development in the U.S. West*, Environmental Challenges and Solutions 4,
DOI 10.1007/978-3-319-53414-5_13

> *will let the free market and the public's preferences determine the industry outcomes. In assessing the various sources of potential energy, Republicans advocate an all-of-the-above diversified approach, taking advantage of all our American God-given resources. That is the best way to advance North American energy independence.* (Republican Party Platform 2016)

> *For decades, America has relied on unsustainable energy sources and has been bound to the status quo by special interests. Democrats have made historic investments in clean-energy technologies that are helping pave the way to a more sustainable future, creating new jobs and entire industries here in America. We believe that now is the moment for this generation to embark on a national mission to unleash America's innovation and seize control of our own destiny. The transition to clean energy has the potential to grow our economy and create millions of new jobs here in America. Recognizing this, the Obama administration has made historic strides in creating a 21st-century clean-energy economy.* (Democratic Party Platform 2016)

The main thrust of this book has been trying to answer the question of whether there are certain shared patterns of beliefs and values held by individuals who either support or oppose the development of alternative energy technologies (AETs). This question is stimulated by several considerations. First, in many ways, AETs can be seen as part of the "green revolution," wherein the goal is to reduce the carbon footprint of humans. Moreover, the frequently articulated goal of sustainability implies a capacity to substitute renewable energy sources for those currently dependent on nonrenewable ones, such as petroleum and coal. Therefore the expectation of this research was that there would be substantial support for AETs among those individuals with environmental values and with the more general political ideologies and values associated with them, alongside opposition to the technologies among those who do not. Such bundles of shared values and ideologies would thus comprise a consideration at play in the development of AETs.

At the same time, however, the movement to develop many AETs faces significant opposition both from those who see no need to employ renewable resources, as well as from many with strong environmentalist values. In the first case, the development of green, renewable energy sources may be seen as unnecessary, either in consideration of criticisms of the data about such phenomena as global warming or about the implications of a rapidly diminishing carbon-based fuel supply. However, perhaps more interesting and less expected is the occasional opposition from environmentalists, especially those with bio-centric (as opposed to anthropocentric) orientations toward the environment. These environmentalists may decry the more local harmful side effects of some AETs, such as wind power, solar power, and wave energy. For example, wind farms may be viewed as a danger to birds, solar farms as hazardous to desert species such as the tortoise, and wave energy farms as potential impediments to migrating whales or bottom-dwelling species.

Finally, there are also concerns by supporters and opponents alike about the processes and institutions that are used in the selection and implementation of AET sites. More specifically, will local socioeconomic, environmental, landscape, and visual concerns be considered by developers and policy makers in the planning and implementation of AET? If local concerns are ignored and planning processes are perceived as either top-down, even if driven by global environmental values, or dominated by untrusted developers and policy makers, public acceptance of AETs may well be damaged, leading to growing challenges for AET developers, planners, and policy makers (Devine-Wright 2011; Haggert 2008; Wolsink 2007, 2009, 2010).

13.2 Process Matters

In a comparative analysis of wind power implementation in the Netherlands, Germany, and England, Breukers and Wolsink concluded (2007, p. 2737): "Facilitating local ownership and institutionalizing participation in project planning can help to arrive at a better recognition and involvement of the multiple interests (environmental, economic and landscape) that are relevant at the local level of implementation." The first two case studies presented here—the failed Calico Solar Project in California and the successfully sited Kittitas Valley Wind Project in Washington—are prime examples of how *not* to approach the siting of AET. Both case studies also provide many lessons for the development and implementation for more nascent AET siting situations, such as wave energy off of the Oregon Coast.

In the case of the Calico Solar Project (Chap. 3), initial stakeholder involvement was triggered by concern about the location of the proposed site and its potential impacts on the desert landscape and wildlife. While the environmental groups that opposed this project are on record as supporting AET, they were concerned about the perceived lack of relevant scientific information available about the impacts of utility-scale solar on wildlife. In addition, stakeholders, including environmental groups, perceived the planning process as unfair and secretive. And, if stakeholders believe the siting planning process is not fair and transparent, future AET siting projects may be put at risk because of damaged relationships and broken channels of communication. Similarly, when local concerns are ignored and siting decisions are the product of top-down processes, such as the Kittitas Valley Wind Project (Chap. 4), local community distrust, anger, and cynicism are the result. While the result of top-down executive decision making in the siting of AET may well lead to distrust in a specific community, it may also make it more difficult to site AET in other communities and make it more difficult for the state in other policy arenas as well (e.g., environmental policies in general).

For the remaining two case studies, presented in Chaps. 5 and 6, there are some lessons to learn for developers, policy makers, and stakeholders alike. While the development of geothermal energy at the Newbury Volcano and wave energy on the Oregon Coast are in testing stages and are undergoing economic viability analyses, there are legitimate public concerns that should be addressed in a transparent and meaningful process. For geothermal development at the Newberry Volcano, there remain public concerns regarding groundwater connectivity, seismicity, toxicity, and other issues, such as the potential impact on the Newberry National Volcanic Monument. A diverse group of stakeholders, including the Klamath Tribe, the Northwest Environmental Defense Center, Oregon Wild, Blue Mountain Biodiversity

Project, and others, are monitoring the project closely and are skeptical that the global environmental benefits of the project outweigh the potential local environmental, geological, social, and economic costs.

The final case study presented—wave energy development on the Oregon Coast—reveals that many stakeholders also concerned about the potential social, economic, and environmental impact of wave buoy farms. According to some observers, wave energy technology is possibly decades behind wind and solar technologies, and is fraught with much uncertainty because of the complexity of the technology (Levitan 2015). As one observer has commented (Levitan 2015), "From a technical point of view, operating in the ocean is far more difficult than on land… Saltwater is a hostile environment for devices, and the waves themselves offer a challenge for energy harvesting as they not only roll past a device but also bob up and down or converge from all sides in confused seas."

This challenge has led to many concerns about the economic costs associated with wave energy and the potential need for large public subsidies to make the technology economically feasible. In addition, fishers and crabbers are concerned about the potential environmental effects on fisheries and the ocean floor that might harm their traditional occupations. There are also aesthetic concerns among many of Oregon's coastal residents about how wave energy may affect tourism and scenic beauty. Oregon has seemingly learned the negative lessons of top-down, non-transparent planning processes by starting slowly with test buoy sites and involving coastal communities in the siting process. This process will need to remain transparent and participatory if and when wave buoy farms become economically viable.

13.3 Environmental Values Matter

The analyses of survey data presented in this book have provided some insight into the question of whether identifiable individual-level attributes (e.g., values, interests, command of relevant scientific knowledge, demographic characteristics) lead to support for or opposition to the development and implementation of AET. The study has identified many such discernable predictors of support and opposition to renewable sources of energy for electrical power. Using public survey data collected during the fall of 2012 and winter of 2013 in California, Oregon, Washington, and Idaho, the analyses examined the role of environmental values in the shaping of public orientations toward environmental impacts versus renewable energy technology tradeoffs, the latter themselves with pro-environmental implications. The findings show that certain shared environmental values are indeed held by sets of individuals who either support or oppose the development of AETs versus a set of hypothetical local environmental conditions. To be sure, those respondents with higher NEP scores were more supportive of local environmental AET considerations when compared to individuals with lower NEP scores, but there also were substantial pro-NEP respondents who were more prone to supporting AET development even if there were site-specific negative environmental consequences.

In terms of state comparisons, not surprisingly Idaho residents were slightly less supportive of the NEP than were residents of the other three "blue" states. Idaho had the lowest overall score on the NEP index. In addition, for the individual items in the index, Idahoans were more likely than were those in the other three states to believe that "the so-called 'ecological crisis' facing humankind has been greatly exaggerated," less likely to believe that "plants and animals have as much right as humans to exist," and more likely to believe that "humans were meant to rule over the rest of nature."

State comparisons of the percentage of the various combinations of AET and NEP measures revealed some small differences between the states as well. Idaho had a slightly higher percentage (18%) of those who were low in environmental orientation on both the AET and the NEP than did California (15%), Washington (16%), and Oregon (14%) residents. In addition, Idaho also had the lowest percentage of "globalists" who were highly environmentally oriented on both measures—pro-AET *and* high NEP. Idaho also had a higher percentage of those who opposed AET projects—even in the lowest category of the NEP—compared to other states. Oregon had the highest percentage of globalists (very high on the NEP, but also very supportive of AET). These results reflect the strong support for AET in the three states with aggressive RPS in place and programs to promote AET, compared to Idaho, which has ample and inexpensive hydropower and much more conservative politics.

As noted, as part of the analyses presented in the survey-based chapters, two distinct value groups were identified, labeled "localists" and "globalists." Localists were those respondents in the top third of the NEP and in the top third of the pro-local environmental conditions on the AET index. Globalists were also in the top third of the NEP, but in the top third of the pro-AET index. A major goal of this research was to determine how and why environmentalists, or those who have strong environmental values, make tradeoff choices between local environmental conditions and global efforts to mitigate greenhouse gas emissions through the use of AET. We found that

- Localists were more likely than globalists to both talk about renewable energy policy issues and to believe that they could have a greater impact on renewable energy policy.
- Globalists were found to be slightly more likely to exhibit postmaterial values, while localists were slightly more likely to have mixed values.
- Globalists were more willing than localists to trade off local negative environmental impacts in order to site alternative energy technologies, were less likely to be very liberal, less likely to believe that the earth's climate is getting warmer, and less likely to believe that change is caused by humans.
- Localists were more likely to be concerned about energy scarcity by perceiving a greater likelihood of insufficient energy resources in the future, compared to globalists.
- Concerning energy policy knowledge, for two knowledge questions (the greatest state source of energy and the greatest state use of energy) there were no

differences between localists and globalists. However, for a third knowledge question (the meaning of "off-the-grid"), localists were more often correct (70%) than were the globalists (57%).
- When compared to globalists, localists were more likely to use the following sources for energy policy information: (1) public radio, (2) major newspapers, (3) local newspapers, (4) community leaders, (4) state elected officials, and (5) environmental groups.

Given that localists were pro-NEP, talked more about energy policy with family and friends, had higher levels of political efficacy, had slightly higher energy knowledge scores, and used more energy information sources, they may be well positioned to oppose local and regional AET sites. Taken together, these are characteristics of those who participate in the political process (Kenski and Stroud 2006; McLeod et al. 1999). While it cannot be determined for sure from these data whether any of the survey respondents have been involved in AET policy issues or siting, given the nature of the questionnaire used, the four case studies did include stakeholders opposed to AET projects because of perceived local and regional environmental harm caused by the projects. Future research should more fully investigate the relationship between AET tradeoffs and the localist-globalist framework.

13.4 Other Things Matter, But Not So Much

Additional analyses presented in this book also examined the impact of energy policy beliefs, energy policy knowledge, and energy policy information-source use on AET tradeoffs. The energy policy beliefs included perceptions of future energy scarcity, the role of energy independence for national security, and the potential ameliorative role of technology to solve our energy supply issues. Prior research suggests that those who believe that the U.S. will have future energy scarcity, who believe national security is jeopardized by dependence on foreign sources, or who are skeptical about the ability of future technology developments to solve energy needs, are most supportive of AET development.

Moreover, large majorities of respondents from each state strongly agreed with the statement, "decreasing our dependence on foreign oil and gas is important to national security and the statement, "I am concerned about foreign ownership of our energy resources." In addition, majorities in each state also agreed and strongly agreed with the statement, "I am concerned that our country doesn't have enough energy resources." Similarly, a majority of respondents in each state agreed that "Not enough money is being spent on research and development of alternative fuels," and "New technologies will make it possible to have enough electricity for all of us in the future." However, while a majority of respondents were concerned about energy security and energy scarcity, but positive about the ability of technology to help solve these perceived problems, how did their energy beliefs impact AET tradeoffs? The analyses presented in Chap. 9 showed mixed results for this question.

There were very weak relationships of scarcity, security, and technology indicators with the tradeoff questions. However, for the wind farm tradeoff question, there was a negative relationship with scarcity and technology. Respondents who saw future energy scarcity as likely were slightly more likely to support the development of a "high potential windmill energy farm in windy, rare tall grass prairie area." In addition, those respondents who were less than optimistic about the ability of technology to solve future energy needs were also more likely to support the development of wind farms, regardless of the local environmental consequences. When combining NEP values and the AET attitudes into globalist and localist types, only the energy policy belief of scarcity distinguished between the two groups, with localists ironically perceiving greater likelihood of insufficient resources.

What are the implications of these findings for the development and siting of AET projects? Framing new projects with public appeals for national security and the ability of future technologies to solve energy needs may not be fruitful for policy makers and developers. However, for the more developed wind energy sector, framing projects as helping to avoid future energy scarcity may be more successful, depending on the technology and project location. Of course, the hypothetical AET tradeoff questions used in the analyses could produce different outcomes if the local environmental tradeoffs involved different scenarios, or if the scenario provided mirrored an actual local project environmental issue, as the case studies have shown. And, even when controlling for energy beliefs, the NEP is still the most important driver of AET tradeoffs.

In regard to the relationship between energy policy knowledge and AET tradeoffs, respondents were asked a series of questions, including a general subjective question concerning self-assessed knowledge of renewable energy technology, and a short quiz on energy policy. For the level of self assessed knowledge, respondents were asked, "In general, how well informed would you consider yourself to be concerning renewable energy policy issues in (state)—such as wind, solar, wave and biomass ("wave" was not included in the Idaho survey). A majority of respondents in each state considered themselves "somewhat informed" and few considered themselves "very well informed." Respondents also were asked about their level of familiarity with individual AETs, including biofuels, wind energy, geothermal energy, solar energy, wave energy, and hydroelectric energy (once again "wave energy" was omitted from the Idaho survey). Respondents in all states expressed familiarity with wind, solar, and hydroelectric energy. Respondents were less familiar with biofuels, geothermal, and wave energy. California respondents were slightly less familiar than respondents from the other states with hydroelectric. This surely reflects the larger percentage of hydroelectric power used in Oregon, Idaho, and Washington energy portfolios, compared to California, where natural gas is more prominent.

The second indicator used to assess respondents' energy knowledge was their ability to correctly answer questions about energy resources in their particular state. Thus, respondents were asked to identify the largest energy source for electricity in their respective states and the economic sector that uses the most energy, and to correctly define the term "being off the grid." Residents of Washington and Idaho were found to more accurately identify the predominant source of energy than were residents of Oregon, and far more likely than those from California. Washington and Oregon residents were also more informed on the largest energy consumption sector than were California and Idaho residents. The percentage of respondents familiar with "being off the grid" was similar across the states.

When examining the independent effects of renewable energy familiarity and knowledge on AET tradeoffs, knowledge had a weak direct effect on only one of the tradeoff scenarios—hydroelectric power. Higher levels of knowledge were associated with greater support for the position to "continue to operate hydroelectric dams for power even if it harms salmon runs." Self-assessed familiarity had a weak impact for two of the tradeoff questions—solar and wind. However, the effects were in contrasting directions, with greater familiarity leading to greater support for the preservation of tall grass prairie "even if it prevents windmill energy farm production," while on the other hand, it also led to somewhat greater support for installing "solar power in desert even if it disrupts ecology of endangered turtle."

The implication of these findings, similar to those concerning energy policy beliefs, is that increasing levels of energy policy knowledge and familiarity may not influence most individuals' choices concerning tradeoffs. However, as with the energy belief questions, other knowledge indicators may yield more robust results than what was used in this study. Once again, while controlling for energy knowledge and familiarity, environmental values were more important in determining AET tradeoffs than were other factors. However, there was an interaction between energy policy knowledge and environmental values—what people know about energy policy does make a difference in how they connect their fundamental environmental values to their positions on the AET tradeoff questions.

The final set of analyses in the book concerned where individuals obtain their information concerning energy policy issues. Survey results found that most respondents obtain their information through television news, followed closely by the Internet, major newspapers, and public broadcasting. California respondents were less likely to use major newspapers, local community leaders, and state elected officials as information sources than were respondents in the other three states. In addition, California and Oregon respondents indicated a greater willingness to use a wider variety of information sources, compared to Washington and Idaho respondents.

There also were some identifiable patterns of information source use. For example, respondents who relied on elected officials for information also relied on local community leaders, state energy departments, and utilities. This group of information sources seems based on notions of political authority and/or expertise in regard to providing information about energy policy. Those respondents who relied on environmental groups for information were also likely to rely on the Internet and universities for information sources; however this group did not rely as much on television news, suggesting a possible preference for alternative and potentially elite-challenging information sources. When analyzing the potential impact of information-source use on AET tradeoffs, those who relied on the Internet and public broadcasting were more likely to oppose the siting of AET because of perceived negative environmental effects. But, once again, environmental values played a role here. Those respondents identifying with the NEP were also most likely to use the Internet and public broadcasting, suggesting that knowledge acquisition is mediated by other values, such as political orientation and environmental values.

When comparing globalists and localists, localists were more likely than globalists to use multiple information sources and to rely more heavily on environmental groups for information. Because localists were more concerned about negative local environmental impacts of AET projects, perhaps seeing a more clearly identifiable self-interest, they were more likely to obtain information about those negative environmental effects from environmental groups. Of course, as prior research suggests, people often seek information from sources that are consistent with their interests and value systems. The implication of these results for promoters of AET projects is that values matter, and that multiple information sources should be used to provide different groups of people with information, so that an open and transparent decision-making process is possible.

13.5 Conclusion

The passages from the 2016 Democratic and Republican Party Platforms in the introduction of this chapter provide quite contrasting views of U.S. energy policy and the role of government in the promotion of certain sources of energy. While government promotion of renewable technologies is prominent in the Democratic Platform, the Republican Platform calls for market-based approaches that promote all forms of energy, including fossil fuels and renewables *if they make economic sense*. The analyses presented here further illustrate the role of values in public preferences concerning AET projects and the environment. Local communities can have legitimate social, economic, and ecological concerns and values concerning proposed AET projects and the processes by which siting decisions are made. As Heberlein in *Navigating Environmental Attitudes* advises planners and policy makers, "technical fixes must be designed to be consistent with attitudes" prevalent in communities (2012, p. 88). And, it should be added, the processes for that design must also be consistent with the values and expectations of the communities. Without proper consideration of such concerns and values, future attempts to implement alternative energy technologies may constitute no more than *tilting at windmills*.

References

Breukers, S., & Wolsink, M. (2007). Wind power implementation in changing institutional landscapes: An international comparison. *Energy Policy, 35*, 2737–2750.

Democratic Party Platform. (2016). *Energy independence*. https://www.democrats.org/issues/energy-independence. Accessed 10 July 2016.

Devise-Wright, P. (2011). Public engagement with large-scale renewable energy technologies: Breaking the cycle of NIMBYism. *WIREs Climate Change, 2*, 19–26.

Haggett, C. (2008). Over the sea and far away? A consideration of the planning, politics and public perception of offshore wind farms. *Journal of Environmental Policy and Planning, 10*, 289–306.

Heberlein, T. A. (2012). *Navigating environmental attitudes*. New York: Oxford University Press.

Kenski, K., & Stroud, N. J. (2006). Connections between internet use and political efficacy, knowledge, and participation. *Journal of Broadcasting & Electronic Media, 50*, 173–192.

Levitan, D. (2015, April 28). Why wave power has lagged far behind as energy source. *Environment 360*. http://e360.yale.edu/content/print.msp?id=2760. Accessed 13 Aug 2016.

McLeod, J. M., Scheufele, D. A., & Moy, P. (1999). Community, communication, and participation: The role of mass media and interpersonal discussion in local political participation. *Political Communication, 16*, 315–336.

Republican Party Platform. (2016). *America's natural resources*. https://www.gop.com/platform/americas-natural-resources/. Accessed 10 July 2016.

Wolsink, M. (2007). Wind power implementation: The nature of public attitudes: equity and fairness instead of 'backyard motives'. *Renewable Energy Sustainability Review, 11*, 1188–1207.

Wolsink, M. (2009). Near-shore wind power-protected seascapes, environmentalists' attitudes, and the technocratic planning perspective. *Land Use Policy, 27*, 195–203.

Wolsink, M. (2010). Contested environmental policy infrastructure: Socio-political acceptance of renewable energy, water, and waste facilities. *Environmental Impact Assessment Review, 30*, 302–311.

Appendices

Appendix A. State Energy Profiles

	California	Idaho	Oregon	Washington
General indicators				
Population (millions) 2013	38.3	1.6	3.9	7.0
Civilian Labor Force (millions) 2014	18.6	0.8	1.9	3.4
Per capita income 2012	$44,980	$33,749	$38,786	$45,413
Employment in Green Jobs (2011) (BLS, BNEF and ACORE, 2013)	360,245	24,250	68,709	101,593
Investment (Grossed-up), 2011/2012	Asset finance: $16.1bn/$8.5bn	Asset finance: $715.7m/$232.3m	Asset finance: $234.6m/$414.8m	Asset finance: $228.8m/$(0)
(BLS, BNEF and ACORE, 2013)	Venture capital & private equity: $1.4bn/$1.6bn	Venture capital & private equity: $(0)/$2.2m	Venture capital & private equity: $100.5m/$42.6m	Venture capital & private equity: $9.5m/$24.9m
Energy consumption & expenditures				
Total energy consumption per capita/U.S. state rank (millions) 2011	209 Btu/47th	332 Btu/21st	262 Btu/39th	305 Btu/29th

(continued)

J.C. Pierce, B.S. Steel, *Prospects for Alternative Energy Development in the U.S. West*, Environmental Challenges and Solutions 4,
DOI 10.1007/978-3-319-53414-5

	California	Idaho	Oregon	Washington
Total energy consumption (trillions Btu) 2011/ U.S Rank (EIA, 2014)	7858/2nd	526/42nd	1014/32nd	2080/16th
Total expenditures/ U.S. rank, 2011 (EIA, 2014)	$ 136,096 m/2nd	$ 6586 m/42nd	$ 14,941 m/31st	$ 27,124/17th
End-use energy consumption by sector (trillions) 2011: (EIA, 2014)				
Residential	1516 Btu	125 Btu	264 Btu	510 Btu
Commercial	1556 Btu	85 Btu	195 Btu	387 Btu
Industrial	1786 Btu	184 Btu	245 Btu	586 Btu
Transportation	3,0001 Btu	132 Btu	310 Btu	598 Btu
End-use energy expenditures by sector (trillions) 2011: (EIA, 2014)				
Residential	$ 19,453 m	$ 1056 m	$ 2636 m	$ 4579 m
Commercial	$ 19,573 m	$ 612 m	$ 1703 m	$ 3058 m
Industrial	$ 14,051 m	$1297 m	$ 1864 m	$ 2922 m
Transportation	$ 83,020 m	$ 3621 m	$ 8738 m	$ 16,567 m
Electricity generation, 2012 (%) (EIA apud ACORA, 2013)				
Coal	1%	<1%	5%	3%
Natural gas	60%	12%	19%	5%
Hydroelectric	13%	73%	65%	77%
Nuclear	9%	0	0	8%
Other gases	1%	0	0	0
Other renewables	16%	15%	11%	7%
Energy consumption estimates (Trillion Btu), 2011 (EIA, 2014)				
Coal	55.3	7.8	35.1	57
Natural gas	2196.3	83.9	203.6	272.3
Motor gasoline excl. Ethanol	1698.1	79.9	169.1	302.9
Distillate fuel	545.4	61	111.1	151
Jet fuel	549.7	3.6	25.5	92.9
LPG	65.2	5.8	6.3	16.8
Residual fuel	186.9	0	7	50.5
Other petroleum	339.6	11.1	20.4	110.5
Nuclear electric power	383.6	0	0	50.3
Hydroelectric power	413.5	130.2	411.1	892.1
Biomass	263.7	34.5	63.3	124.3
Other renewables	259.7	14.9	50.4	62.5
Net interstate flow of electricity	868.6	93	−90	−79.9

(continued)

	California	Idaho	Oregon	Washington
Energy production estimates (Trillion Btu), 2011 (EIA, 2014)				
Coal	0	0	0	0
Natural gas/ marketed	279.7	0	0	0
Crude oil	1123.4	0	0	0
Nuclear electric power	383.6	0	0	50.3
Biofuels	25	7.6	5.7	0
Other renewable energy	812.8	172.8	508	1051.6
Installed renewable energy capacity, 2012 (ACORE, 2013)				
Wind power	5544 MW	973 MW	3153 MW	2808 MW
Solar photovoltaic	2559.3 MW	1 MW	56.4 MW	19.5 MW
Solar thermal electric	393.2 MW	0 MW	0 MW	0 MW
Geothermal power	2732.2 MW	15.8 MW	33.3 MW	0 MW
Hydropower	10,053.6 MW	2536.2 MW	8241.4 MW	20,902.5 MW
Biomass power	1416.8 MW	148.1 MW	402.6 MW	403.3 MW
Ethanol	223 mGy	50 mGy	149 mGy	0 mGy
Biodiesel	78 mGy	6 mGy	18 mGy	117 mGy
Total installed renewable energy capacity, 2012 (ACORE, 2013)	22,699 MW; 301 mGy	3674 MW; 56 mGy	11,887 MW; 167 mGy	24,133 MW; 117 mGy

Appendix B

California, Idaho, Oregon, and Washington Energy Policy Survey

OSU
Oregon State
UNIVERSITY

Please return surveys to:

Energy Policy Survey
School of Public Policy
300 Gilkey Hall
Oregon State University
Corvallis, Oregon 97331-6206
541-737-2811

ID # ___________________

[for mailing purposes only]

SECTION 1

In this first section of the survey we would like to ask you some general questions about your interest, activities, and knowledge about energy and environmental policy issues. Please circle the number that most closely represents your view.

Q-1 In general, how well informed would you consider yourself to be concerning renewable energy policy issues in Oregon—such as wind, solar, wave, and biomass energy?

1. Not informed
2. Somewhat informed
3. Informed
4. Very well informed

Q-2 How often do you talk about issues related to renewable energy with your family, friends, or other acquaintances?

1. Never
2. Hardly ever
3. Sometimes
4. Often

Q-3 Overall, how much impact do you think PEOPLE LIKE YOU can have in making Oregon's energy policy?

1. No impact at all
2. A small impact
3. A moderate impact
4. A big impact

Q-4 How familiar are you with specific renewable energy technologies including biofuel, wind, solar, geothermal and wave energy? Please circle the number of your response.

	Not Familiar	**Somewhat Familiar**	**Familiar**	**Very Familiar**
a. Biofuels (e.g. ethanol, etc.)	1	2	3	4
b. Wind energy	1	2	3	4
c. Geothermal energy	1	2	3	4
d. Solar energy	1	2	3	4
e. Wave energy	1	2	3	4
f. Hydroelectric energy	1	2	3	4

Q-5 How much do you agree or disagree with the following statements concerning energy policy?

	Strongly Disagree	Somewhat Disagree	Neutral	Somewhat Agree	Strongly Agree
a. Decreasing our dependence on foreign oil and gas is important to our national security.	1	2	3	4	5
b. I am concerned about foreign ownership of our energy resources.	1	2	3	4	5
c. I am concerned that our country doesn't have enough energy resources.	1	2	3	4	5
d. I am concerned about being personally affected by shortage of electricity in the next five years.	1	2	3	4	5
e. Not enough money is being spent on research and development of alternative fuels.	1	2	3	4	5
f. It is possible to increase energy supplies while protecting the environment at the same time.	1	2	3	4	5
g. New technologies will make it possible to have enough electricity for all of us in the future.	1	2	3	4	5

Q-6 Here are a few specific questions about energy. Many people don't know the answers to these questions, so if there are some you don't know just leave them blank and continue.

a. The largest source of energy for electricity in Oregon is:

1. Coal
2. Hydroelectric
3. Natural Gas
4. Nuclear

b. Most electricity in Oregon is used by the:

1. Residential Sector (e.g., households)
2. Commercial Sector (e.g., retail stores)
3. Industrial Sector (e.g., factories and mills)
4. Transportation Sector

c. Being "off-grid" means:

1. Producing one's own electricity
2. Getting electricity from another state
3. Having no electricity
4. Being energy efficient

Q-7 Wave energy refers to the extraction of electricity from the up-and-down motion of ocean waves using buoys or devices in the form of "wave energy farms." What is your general attitude toward the development of wave energy off of the Oregon coast?

1. Very positive
2. Positive
3. Neutral
4. Negative
5. Very negative
6. Do not have enough information to form an opinion.

Q-8 In recent years there has been considerable attention and development of what many call *alternative energy technologies,* such as **solar power**, **wind power**, **wave and tidal energy** and **biofuels**. Sometimes, though, alternative energy technologies have impacts that some people find undesirable. We are interested in ***your views*** of what should be done in cases where those impacts are likely to occur. Please circle the number in each case that best represents your opinion. [1=Strongly Agree; 2=Agree; 3=Neutral; 4=Agree; 5=Strongly Agree]

a. Develop high potential windmill energy "farm" in windy rare "tall grass prairie" area. — Preserve rare "tall grass prairie" ecology even if this prevents windmill energy farm production.

1 2 3 (neutral) 4 5

b. Leave undisturbed desert floor ecology that supports endangered turtle species even if it prevents solar power installation. — Install solar power farm in desert even if it disrupts ecology of endangered turtle.

1 2 3 (neutral) 4 5

c. Install off-shore wave turbines on the sea floor to produce electricity even if it impedes whale migration. — Protect whale migratory routes even if it means prohibiting installation of off-shore wave energy turbines.

1 2 3 (neutral) 4 5

d. Protect traditional native/tribal cultural sites even if it would prevent building transmission lines or roads for alternative energy (solar, wind, biofuel) supplies. — Build alternative energy transmission lines even if it would disturb traditional tribal/native cultural sites.

1 2 3 (neutral) 4 5

e. Continue to operate hydroelectric dams for power even if it harms salmon runs. — Remove dams where they cause harm to salmon runs.

1 2 3 (neutral) 4 5

Q-9 We would like to know which of the following information sources you currently use or would use to learn more about California's energy situation and policy. Please circle the number of the frequency of your use.

	Never	Infrequently	Frequently	Very Frequently
a. Television news programs and specials	1	2	3	4
b. Public Broadcasting	1	2	3	4
c. Radio programs	1	2	3	4
d. Major newspaper	1	2	3	4
e. Other local newspapers	1	2	3	4
f. Local community leaders	1	2	3	4
g. State elected officials	1	2	3	4
h. State Department of Energy	1	2	3	4
i. Universities and colleges	1	2	3	4
j. Utilities	1	2	3	4
k. Environmental groups	1	2	3	4
l. Information available on the Internet	1	2	3	4
m. Other? ______ (please list)	1	2	3	4

SECTION 2

This section of the survey concerns your attitudes toward the environment and politics. Please circle the number that most closely represents your view.

Q-10 Listed below are statements about the relationship between humans and the environment. For each, please indicate your level of agreement.

	Strongly Disagree	Mildly Disagree	Neutral	Mildly Agree	Strongly Agree
a. The balance of nature is very delicate and easily upset by human activities.	1	2	3	4	5
b. Humans have the right to modify the natural environment to suit their needs.	1	2	3	4	5
c. We are approaching the limit of people the earth can support.	1	2	3	4	5
d. The so-called "ecological crisis" facing humankind has been greatly exaggerated.	1	2	3	4	5
e. Plants and animals have as much right as humans to exist.	1	2	3	4	5
f. Humans were meant to rule over the rest of nature	1	2	3	4	5

Q-11 From what you've read and heard, is there solid evidence that the average temperature on earth has been getting warmer over the past few decades, or not?

1. Yes (go to Q-12)
2. No (go to Q-13)
3. Don't know (go to Q-13)

Q-12 Do you believe that the earth is getting warmer…?

1. Mostly because of human activity such as burning fossil fuels.
2. Mostly because of natural patterns in the earth's environment.
3. Don't know

Q-13 On domestic policy issues, would you consider yourself to be?

1. Very Liberal 2. Liberal 3. Moderate 4. Conservative 5. Very Conservative

Q-14 There is a lot of talk these days about what our country's goals should be for the next ten or fifteen years. Listed below are some goals that different people say should be given top priority. Please mark the one you consider the most important in the long run. What would be your second choice? Please mark that second choice as well.

	1st Choice (circle one)	2nd Choice (circle one)
a. Maintaining order in the nation.	1	1
b. Giving people more say in important governmental decisions.	2	2
c. Fighting rising prices.	3	3
d. Protecting freedom of speech.	4	4

SECTION 3

We now have a few concluding questions to check if our survey is representative of all types of people. Please remember that all answers are completely confidential to the extent permitted by law.

Q-15 What is your current age in years ____________ ?

Q-16 Please indicate your gender? 1. Female 2. Male

Q-17 What level of education have you completed?

1. Grade School
2. Middle or junior high school
3. High school
4. Vocational school
5. Some college
6. College graduate
7. Graduate school
8. Other ____________________________ ?

Q-18 How long have you lived in Oregon? ____________(in years)

Q-19 Which of the following best describes your current work situation?

1. Employed full time
2. Employed part time
3. Not employed outside the home
4. Unemployed
5. Retired
6. Other ____________________________ ?

Q-20 Which category best describes your household income (before taxes) in 2011?

1. Less than $10,000
2. $10,000 - $ 14,999
3. $15,000 - $24,999
4. $25,000 - $34,999
5. $35,000 - $49,999
6. $50,000 - $74,999
7. $75,000 – $99,999
8. $100,000 – $149,999
9. $150,000 – $199,999
10. $200,000 or more

Those are all the questions we have. If you have any additional comments, please include those on a separate piece of paper. Thank you for your precious time.

Index

J.C. Pierce, B.S. Steel, *Prospects for Alternative Energy Development in the U.S. West*, Environmental Challenges and Solutions 4,
DOI 10.1007/978-3-319-53414-5

CPSIA information can be obtained
at www.ICGtesting.com
Printed in the USA
LVOW10*0317230517
535430LV00008B/226/P

9 783319 534138